Ego (

mea serviemus Domino.

Automated Time Series Forecasting Made Easy with R

An intuitive Step by Step Introduction for Data Science

Dr. N.D Lewis

Ordering Information: Quantity sales. Special discounts are available on quantity purchases by corporations, associations, and others. For details, email: info@NigelDLewis.com

Image photography by Deanna Lewis with helpful assistance from Naomi Lewis.

ISBN: 978-1548839017
ISBN: 1548839019

Contents

Dedicated to Angela, wife, friend and mother extraordinaire.

Acknowledgments

A special thank you to:

My wife Angela, for her patience and constant encouragement.

My daughters Deanna and Naomi, for being helpful, taking hundreds of photographs for this book and my website.

And the readers of my earlier books who contacted me with questions and suggestions.

Finally, A Blueprint for Automated Time Series Forecasting with R!

Automated Time Series Forecasting Made Easy with R offers a practical tutorial that uses hands-on examples to step through real-world applications using clear and practical case studies. Through this process it takes you on a gentle, fun and unhurried journey to creating your own models to forecast time series data. Whether you are new to time series forecasting or a veteran, this book offers a powerful set of tools for quickly and easily gaining insight from your data using R.

NO EXPERIENCE REQUIRED: Through a simple to follow step by step process you will learn how to build time series forecasting models using R. Once you have mastered the process, it will be easy for you to translate your knowledge into your own powerful applications.

THIS BOOK IS FOR YOU IF YOU WANT:

- Explanations rather than mathematical derivation.
- Practical illustrations that use real data.
- Worked examples in R you can easily follow and immediately implement.
- Ideas you can actually use and try out with your own data.

TAKE THE SHORTCUT: **Automated Time Series Forecasting Made Easy with R** was written for people who want to get up to speed as quickly as possible. In this book you will learn how to:

- **Unleash** the power of Facebook's Prophet forecasting algorithm.

- **Master** the winning Theta method and use it to predict future observations.
- **Develop** hands on skills using the component form exponential smoothing framework.
- **Design** successful applications using classical ARIMA modeling.
- **Adapt** the flexible BATS and TBATS framework for optimum success.
- **Deploy** the multiple aggregation prediction algorithm.
- **Explore** the potential of simple moving averages.

QUICK AND EASY: For each time series forecasting technique, every step in the process is detailed, from preparing the data for analysis to evaluating the results. These steps will build the knowledge you need to apply them to your own data science tasks. Using plain language, this book offers a simple, intuitive, practical, non-mathematical, easy to follow guide to the most successful ideas, outstanding techniques and usable solutions available using R.

GET STARTED TODAY! Everything you need to get started is contained within this book. **Automated Time Series Forecasting Made Easy with R** is your very own hands on practical, tactical, easy to follow guide to mastery.

Buy this book today and accelerate your progress!

Other Books by N.D Lewis

- Neural Networks for Time Series Forecasting with R
- Machine Learning Made Easy with R
- Deep Learning Made Easy with R:
  - Volume I: A Gentle Introduction for Data Science
  - Volume II: Practical Tools for Data Science
  - Volume III: Breakthrough Techniques to Transform Performance
- Deep Learning for Business with R
- Build Your Own Neural Network TODAY!
- 92 Applied Predictive Modeling Techniques in R
- 100 Statistical Tests in R
- Visualizing Complex Data Using R
- Learning from Data Made Easy with R
- Deep Time Series Forecasting with Python
- Deep Learning for Business with Python
- Deep Learning Step by Step with Python

For further detail's visit www.AusCov.com

Preface

THIS book is about understanding and then hands on use of time series forecasting techniques for prediction; more precisely, it is an attempt to give you the tools you need to build your own time series forecasting models easily and quickly using R. The objective is to provide you the reader with the necessary knowledge to do the job, and provide sufficient illustrations to make you think about genuine applications in your own field of interest. I hope the process is not only beneficial but enjoyable.

On its own, this text won't turn you into a time series forecasting guru any more than a few dance lessons will turn you into the principal dancer with the Royal Ballet in London. But if you're a working professional, economist, business analyst or just interested in trying popular time series models you will learn the basics, and get to play with some cool tools. Once you have mastered the fundamentals, you will be able to apply these ideas using your own data.

Caution!

If you are looking for detailed mathematical derivations, lemmas, proofs or implementation tips, please do not purchase this book. It contains none of those things. This text is designed to be easy to read, simple to understand, and gives straightforward guidance on how to implement the ideas using R.

You don't need to know complex mathematics, algorithms or object-oriented programming to use this text. It skips all that stuff and concentrates on sharing code, examples and illustrations that gets practical stuff done.

Before you buy this book, ask yourself the following tough questions. Are you willing to invest the time, and then work through the examples and illustrations required to take your knowledge to the next level? If the answer is yes, then by all means click that buy button so I can purchase my next cappuccino.

A Promise

No matter who you are, no matter where you are from, no matter your background or schooling, you have the ability to master the ideas outlined in this book. With the appropriate software tool, a little persistence and the right guide, I personally believe the techniques outlined in this book can be successfully used in the hands of anyone who has a real interest.

When you are done with this book, you will be able to implement one or more of the ideas I've talked about in your own particular area of interest. You will be amazed at how quick and easy the techniques are to develop and test. With only a few different uses you will soon become a skilled practitioner.

I invite you therefore to put what you read in these pages into action. To help you do that, I've created **"12 Resources to Supercharge Your Productivity in R"**, it is yours for **FREE**. Simply go to *http://www.AusCov.com* and download it now. It is my gift to you. It shares with you 12 of the very best resources you can use to boost your productivity in R.

Now, it's your turn!

Dr. N.D. Lewis

Chapter 1

Introduction to Time Series Analysis with R

TIME series forecasting has faced a resurgent interest in the past few years. Part of the explanation lies in their continual improvement due to the enhancements in computing power. Another part of the reason is the rapid growth in the availability of data as a result of the explosion of the Internet and digital commerce. Your computer tracks websites visited, social media collects data on your interests and friendships; even your smartphone knows your location and favorite route to work. Retailers, corporations, pharmaceutical companies, local government and national governments are collecting, storing and analyzing more data than ever. Much of this data is suitable for time series prediction.

This introductory chapter:

- Outlines who this book is for.
- Introduces time series data and discusses some of its characteristics.
- Explains how you can get the most out of this book.
- Provides information on getting and using R.

Buried deep in the ever-growing mountain of electronic digits are complex relationships, new insights and groundbreaking discoveries. Time series forecasting tools are designed to help people like you extract this knowledge.

Who is this Book for?

It's no accident that the words simple, easy and gentle appear so often in this text. I have shelves filled with books about forecasting and prediction, including dry statistics texts, dusty computer science manuals, and barely intelligible econometrics articles. Some are excellent, others are good, or at least useful enough to keep. But they range from very long to the very mathematical. I believe many working professionals want something short, simple with practical examples that are easy to follow and straightforward to implement. In short, a very gentle intuitive introduction to time series analysis with R.

The largest users of time series forecasting are in industry and business, yet almost all advice comes from academics; this comes from a practitioner. I have been a practitioner for most of my working life. I enjoy boiling down complex ideas and techniques into applied, simple and easy to understand language that works. Why spend five hours ploughing through technical equations, proofs and lemmas when the core idea can be explained in ten minutes and deployed in fifteen?

I wrote this book because I don't want you to spend your time struggling with the mechanics of implementation or theoretical details. That's why we have Ivy league (in the US) or Russell Group (in the UK) professors. Even if you've never attempted to forecast anything, you can easily make your computer do the grunt work. This book will teach you how to apply the very best time series forecasting tools using R.

How to Get the Absolute Most Possible Benefit from this Book

I want you to get the absolute most possible benefit from this book in the minimum amount of time. You can achieve this by typing in the examples and most importantly experimenting. This book will deliver the most value to you if you do this.

Successfully applying time series algorithms requires work, patience, diligence and most importantly experimentation and testing. By working through the examples and reading the documentation associated with the packages introduced in each chapter, you will broaden your knowledge, deepen your intuitive understanding and strengthen your practical skill set.

Getting R

R is a free software environment for statistical computing and graphics. It is available for all the major operating systems. Due to the popularity of Windows, examples in this book use the Windows version of R. You can download a copy of R from the R Project for Statistical Computing.

This text is not a tutorial on using R. However, as you work through the examples you will no doubt pick up many tips and tricks. If you are new to R, or have not used it in a while, refresh your memory by reading the amazing free tutorials at `http://cran.r-project.org/other-docs.html`. You will be "up to speed" in record time!

Using Packages

If a package mentioned in the text is not installed on your machine you can download it by typing `install.packages("package_name")`. For example, to download the `fpc` package you would type in the R console:

```
install.packages("fpc")
```

Once the package is installed, you must call it. You do this by typing in the R console:

```
require(fpc)
```

The `fpc` package is now ready for use. You only need type this once, at the start of your R session.

> ***NOTE...*** ✍
>
> If you are using Windows you can easily upgrade to the latest version of R using the `installr` package. Enter the following:
>
> ```
> install.packages("installr")
> installr::updateR()
> ```

Effective Use of Functions

Functions in R often have multiple parameters. The examples in this text focus primarily on the key parameters required for rapid model development. For information on additional parameters available in a function, type in the R console: `?function_name`; for example, to find out about additional parameters in the `prcomp` function, you would type:

```
?prcomp
```

Details of the function and additional parameters will appear in your default web browser. After fitting your model of interest, you are strongly encouraged to experiment with additional parameters.

I have also included the `set.seed` method in the R code samples throughout this text to assist you in reproducing the results exactly as they appear on the page.

NOTE... ✍

Can't remember what you typed two hours ago! Don't worry, neither can I! Provided you are logged into the same R session you simply need to type:

```
history(Inf)
```

It will return your entire history of entered commands for your current session.

R User Groups

R user groups are popping up everywhere. Look for one in your local town or city. Join it! Here are a few resources to get you started:

- For my fellow Londoners check out: `http://www.londonr.org/`.
- A global directory is listed at: `http://blog.revolutionanalytics.com/local-r-groups.html`.
- Another global directory is available at: `http://r-users-group.meetup.com/`.
- Keep in touch and up to date with useful information in my FREE newsletter. Sign up at `www.AusCov.Com`.

Characteristics of Time Series Data

Time series is a discrete or continuous sequence of observations that depend on time. Time is an important feature in natural processes such as air temperature, pulse of the heart, or waves crashing on a sandy beach. It is also important in many business processes such as the total units of a newly released book sold in the first 30 days, or the number of calls received by a customer service center over a holiday weekend.

Time is a natural element that is always present when data is collected. Time series analysis involves working with time based data in order to make predictions about the future. The time period may be measured in years, seasons, months, days, hours, minutes, seconds or any other suitable unit of time.

Serial dependence is the key feature of a time series, this means that an observation today is potentially influenced by previous observations (or auto-correlated). To take advantage of serial dependence, time series models use past observations as the basis for predicting future values.

Business and Economic Time Series

Throughout the year, in every economy across the globe, business time series data is collected to record economic activity, assist management make decisions, and support policy-makers select an appropriate course of action.

Figure 1.1 shows the monthly price of chicken (whole bird) from 2001 to 2015. It is characteristic of many business and economic time series. Over the sample period, the price shows a marked upward trend, peaking at over 110 cents per pound. It also exhibits quite marked dips, and periods of increased volatility.

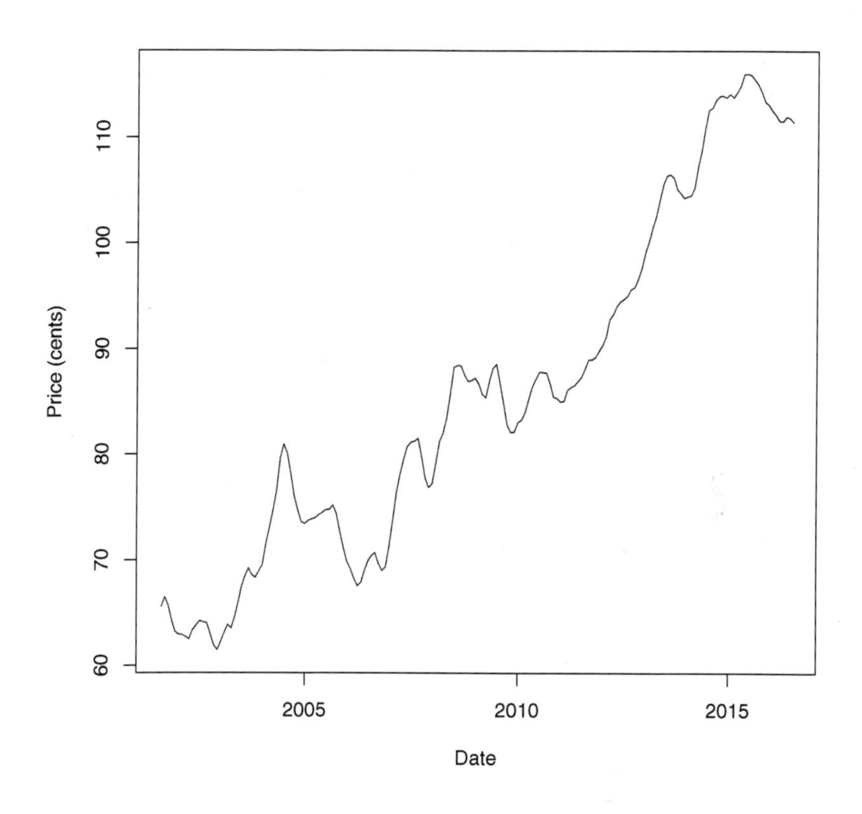

Figure 1.1: Monthly price of chicken

Health and Medical Time Series

Time Series data are routinely collected in medical and health related activities. Blood pressure, weight, heart rate, and numerous other metrics are recorded, often continuously, by various medical devices.

Figure 1.2 shows an electrocardiogram (ECG). An ECG measures the electrical activity of the heart muscle as it changes through time. It is characterized by an undulating cycle with sharp peaks and troughs.

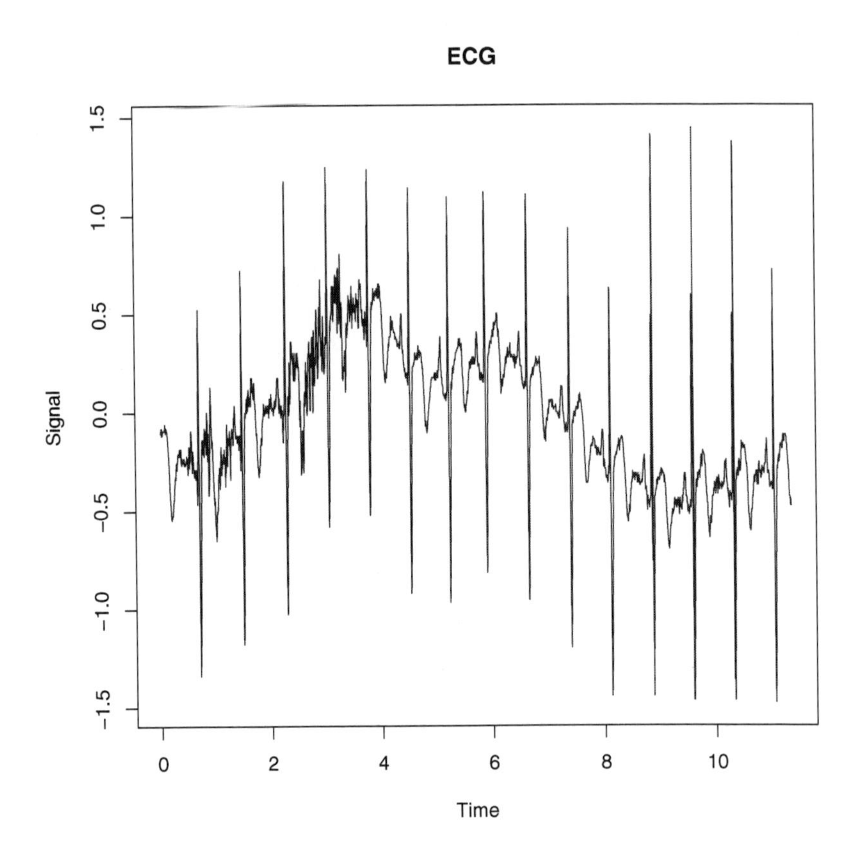

Figure 1.2: Electrocardiogram activity

Physical and Environmental Time Series

The physical and environmental sciences also collect and analyze large amounts of time series data. Environmental variables, collected over long periods of time, help assist in determining climatic trends and future possibilities.

Figure 1.3 shows a time-series plot of the average monthly temperature in Nottingham, England. There does not appear to be any trend evident in the data, however it does exhibit strong seasonality, with peak temperatures occurring in the summer months and lows during the winter.

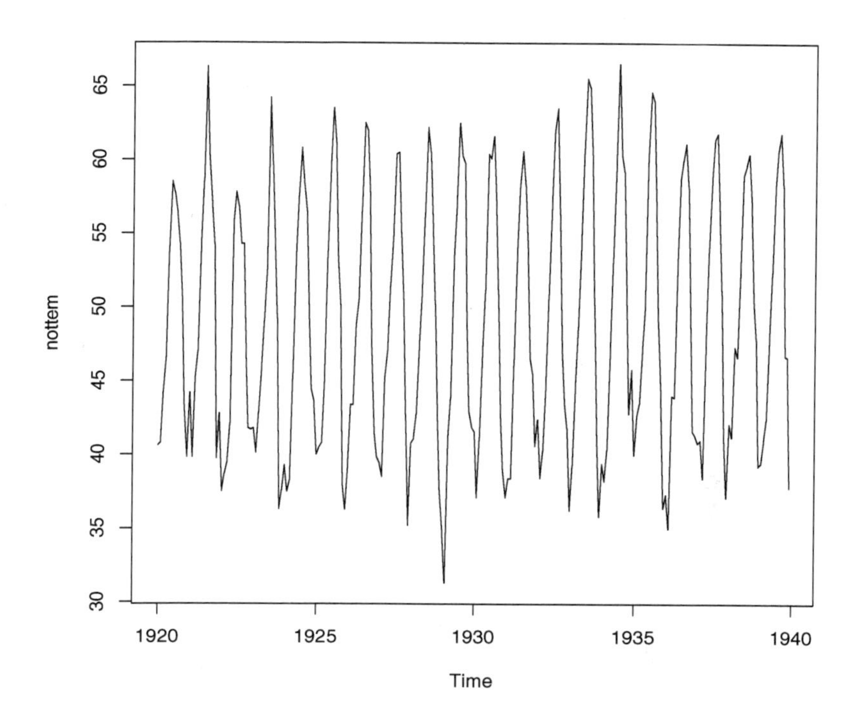

Figure 1.3: Average Monthly Temperatures at Nottingham 1920–1939

Time Series of Counts

Frequently, a time series is composed of counts through time. For example, public health officials in Germany count the number of E.coli cases occurring in a region.

Figure 1.4 shows the weekly count of E.coli cases in the state of North Rhine- Westphalia from January 2001 to May 2013. It appears the number of cases took a dramatic upward turn in late 2011. Since we do not have any information about the cause of the spike, our time series model will have to be flexible enough to handle such upward (or downward) jumps.

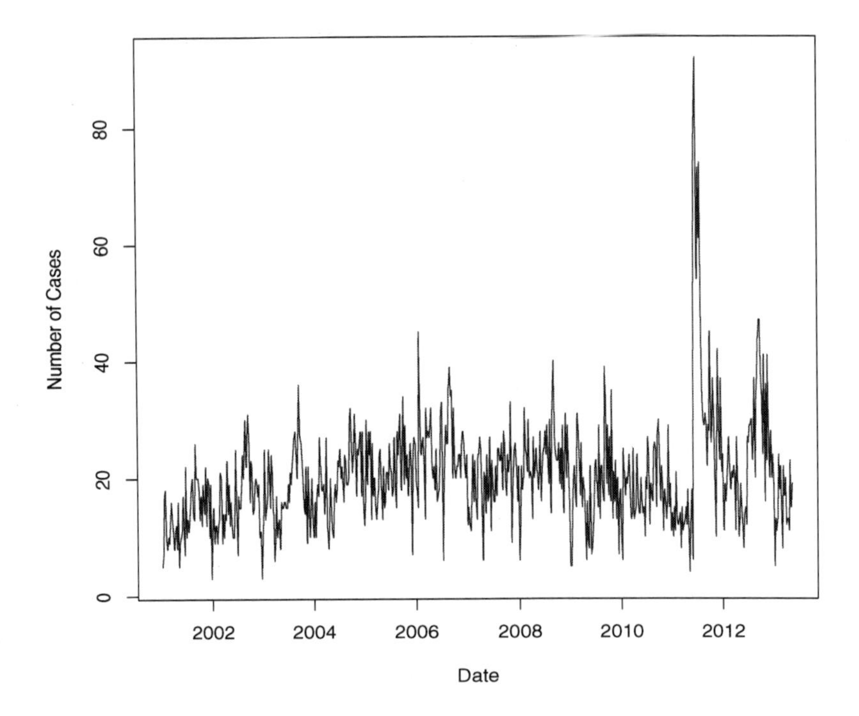

Figure 1.4: Weekly number of Cases of E. coli infection in the state of North Rhine- Westphalia

The Data Generating Mechanism

Time series data are the output of a "*data generating process*". As illustrated in Figure 1.5, at each point in time a new observation is generated. So, for example, at time t we might observe an observation say y. We denote this by y_t. At the next time step, say $t+1$, we observe a new observation, which we denote y_{t+1}. The time step t, might be measured in seconds, hours, days, weeks, months, years and so on. For example, stock market volatility is calculated daily, and the unemployment rate reported monthly.

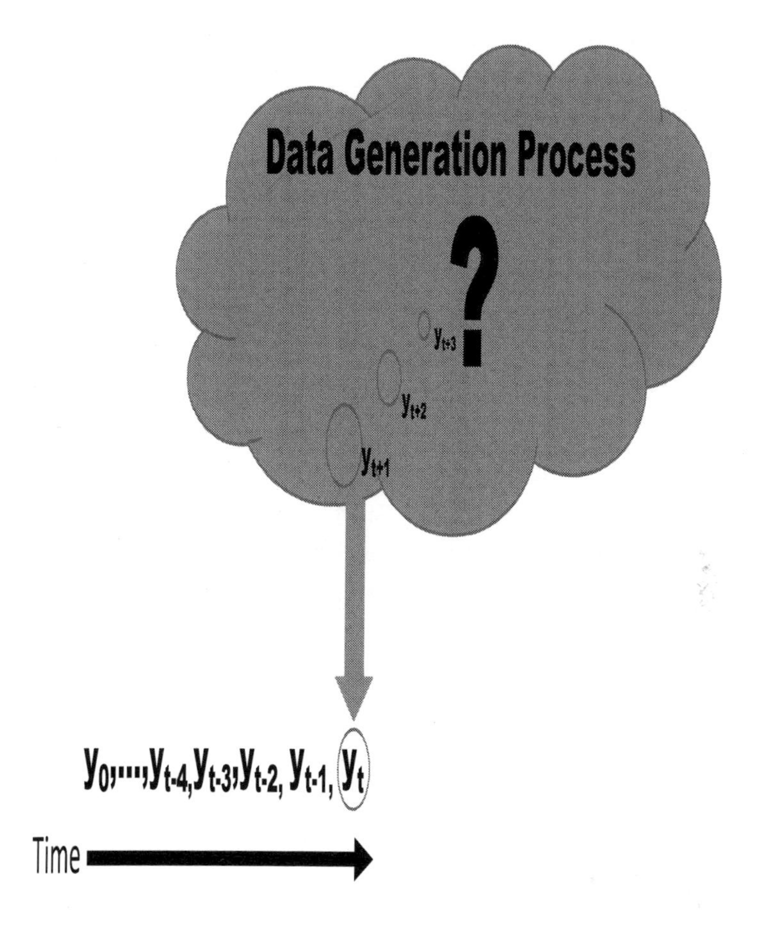

Figure 1.5: Data generating mechanism for time series data

Time Series Decomposition

The mechanism generating time series y_t is often thought to consist of three components. A seasonal component (S_t), a trend component (T_t) , and a random error or remainder component (ϵ_t). Provided seasonal variation around the trend cycle does not vary with the level of the time series, we might write:

$$y_t = S_t + T_t + \epsilon_t$$

This is known as the additive model.

An Example

Take another look at Figure 1.3, it might be a good candidate for the additive model. In R you can use the `decompose` function to estimate each of the components:

```
data("nottem",package="datasets")
y<- nottem
y_dec<-decompose(y,  type = "additive")
plot(y_dec)
```

Here is how to read the above lines of R code. The first line loads the data. It is held in the R object `nottem` from the `datasets` package. The data is transferred to the R object y, and in the third line passed to the `decompose` function. The argument `type = "additive"` tells R to use the additive model. The result is stored in the R object `y_dec`, which is visualized using the `plot` function.

Figure 1.6 shows the original (top panel) and decomposed components (bottom three panels). The components can be added together to reconstruct the original time series shown in the top panel. Notice that:

1. The trend is nonlinear;
2. The seasonal component appears to be constant, so all the years in the sample have a very similar pattern;
3. The random component, shown in the bottom panel, shows no clear trend or cycles.

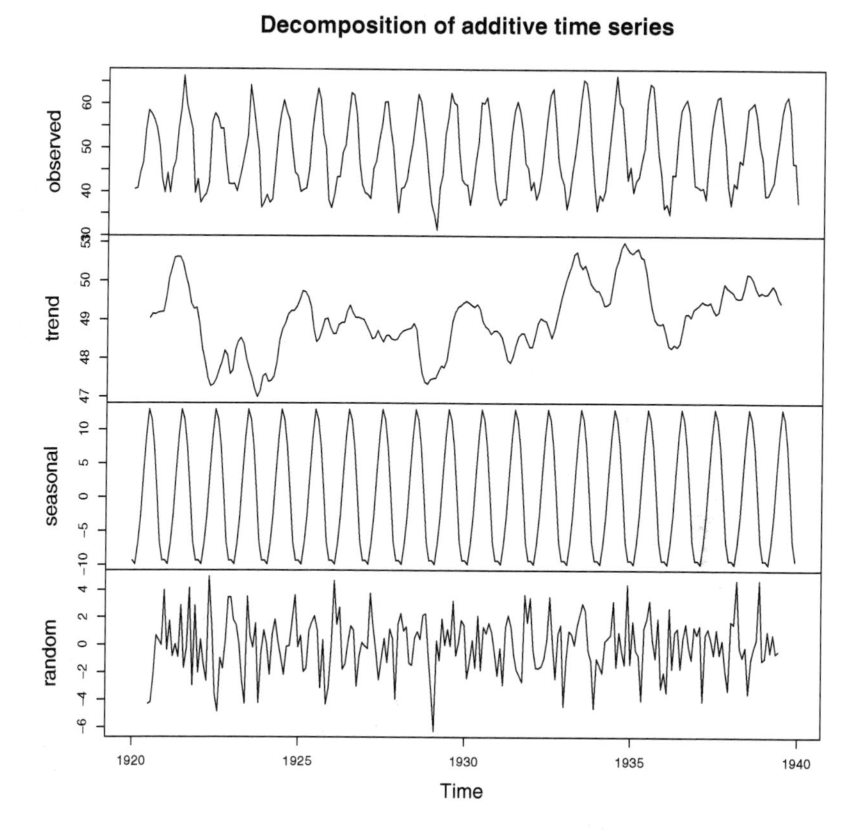

Figure 1.6: Additive model decomposition of the `nottem` dataframe

The multiplicative model

If the seasonal variation around the trend cycle varies with the level of the time series, the multiplicative decomposition can be specified as:

$$y_t = S_t \times T_t \times \epsilon_t$$

NOTE... ✍

You can access the multiplicative model in R by setting `type = "multiplicative"` in the `decompose` function.

What is Automatic Forecasting?

In business, it is common to have hundreds or even thousands of product lines for which a forecast is required on a daily, weekly, monthly or even quarterly basis. This book focuses on time series models that can be built with the minimum amount of "fuss" to help with this issue. By "fuss", we mean first that the models discussed in this text can be built easily and quickly without being hampered by a morass of theoretical constraints.

Second, in very many circumstances time series forecasting requires advanced mathematical and statistical knowledge and there may be nobody suitably trained. In practice, forecasts are generated at all levels of a business organization, by individuals with diverse backgrounds and skill-sets. The ideas discussed in this text can be quickly deployed by almost anyone in the minimum amount of time.

Finally, by “no fuss” we mean that the techniques can be used to generate automatic forecasts of large numbers of univariate time series with the minimum amount of analyst intervention.

Summary

Time-series data consists of sampled data points taken over time from a generally unknown process (see Figure 1.5). It has been called one of the top ten challenging problems in predictive analytics.

I have found, over and over, that an individual who has exposure to a broad range of modeling tools and applications will run circles around the narrowly focused genius who has only been exposed to the tools of their particular discipline. Knowledge of how to build and apply time series models using R will add considerably to your own personal toolkit.

The master painter Vincent van Gough once said

> "*Great things are not done by impulse, but by a series of small things brought together.*"

In the remaining chapters of this text, we will apply this idea to help us develop an understanding of a number of powerful and highly successful time series techniques. Now let's get started!

> ***NOTE...*** ✍
>
> As you use the ideas in this book successfully in your own area of expertise, write and let me know. I'd love to hear from you. Email or visit `www.AusCov.com`.

Chapter 2

Maximizing Use of Simple Moving Averages

MOVING averages are one of the most widely used statistical techniques in business to identify trends in data. They are easy to understand, simple to use, and can be an extremely consistent and reliable means of projecting a trend out into the future. For example, moving averages are frequently deployed by sales managers, who often use a three-month moving average of sales data; and they are used to assess financial data, such as stock market returns, and in economics to estimate trends in macroeconomic data such as unemployment, or interest rates.

By the end of this chapter you will understand:

- The fundamental approach of moving averages.
- The impact of model order.
- Two common performance metrics.
- How to build and automate the construction of moving average models using R.

The field of time series forecasting provides a wide set of techniques for predicting future values. The Simple Moving Average model (SMA) uses smoothed historical data to make forecasts for future periods. Keep reading to see how easy it is to use R to start forccasting using real-world data. Now let's get started!

Understanding Moving Averages

Moving averages are one of the most popular techniques used to project the trend in a time series. There are many variants of moving average. The simplest uses the arithmetic 'mean' or average of past observations as your forecast. This is constructed by summing the observations over a given time period (i.e. say m) and then dividing by m. The parameter m is often called the order of the moving average.

For example, to construct a forecast using a ten-day average, take the last ten daily observations, add them up and divide by ten; the answer gives your average, which is used as the forecast. Tomorrow you would recalculate the average leaving out the price eleven days ago. This approach is called the *simple moving average* time series forecast model or SMA. Its equation for predicting a future observation at time $t + 1$ is given by:

$$\hat{y}_{t+1} = \frac{(y_{t-1} + y_{t-2} + ... + y_{t-m+1})}{m} \tag{2.1}$$

where $\hat{y}_{t+1}$ is the forecast value for the next time period, y the observed values of the time series, and m the order or number of terms used to calculate the average.

Future values are calculated using a window containing the most recent m observations. All observations are treated equally so that each value has an equal weight of $\frac{1}{m}$ in the formula. The forecast is then the arithmetic mean calculated from these observations.

As time passes, the oldest observation is removed from the calculation and the latest value added. This allows the forecast to adjust to the dynamics of the time series over time.

NOTE...

Moving averages are lagging indicators. The larger the order m, the larger the lag.

Clarifying Window Size

The parameter m controls how many past observations are used to estimate the average value. These observations form a "window". If the window is set so that $m = 3$ then:

- The first observation (y_{t-1}) is "*one period old*" relative to the point in time for which the forecast is being calculated;
- the second observation (y_{t-2}) is "*two periods old*";
- and the third observation (y_{t-3}) is "*three periods old*". Hence, the "average age" of the data in the forecast is:

$$\frac{(1+2+3)}{3} = 2$$

This gives an indication of the degree by which the forecast will tend to lag behind the current observations. In the above calculation, the average age was 2, so that is the number of periods by which forecasts will tend to lag behind the current observation.

NOTE... ✍

The rate of response of a SMA model to changes in the observed data is dependent on the number of periods or order, m, included in the moving average equation 2.1.

Impact of different values of m

Figure 2.1 shows the monthly patient count for a medical product. It is characteristic of many such time series in that the observations fluctuate (for the most part) in a defined range. In this example the range is between 700 to 775 units per month. However, there appears to be a trend downwards from 2000 to 2002, and a strong upward trend peaking at the end of 2006. The precise nature of the trend is obscured somewhat by the high level of volatility in the number of units sold.

We can get a better sense of the trend in the data by smoothing it using moving averages. Figure 2.2 illustrates the smoothed data using various values of m. You can see that smaller values of m allow the moving average values to closely follow short term fluctuations of the time series. If m is large the average moves more slowly, producing a much smoother curve. This can be seen more clearly if you compare the diagram $m = 2$, to the diagram where $m = 12$.

The key point is that larger values of m filter out more of the period-to-period noise, and yield a smoother-looking series of forecasts. Smaller values allow the forecast to respond quickly to more recent changes in the underlying process.

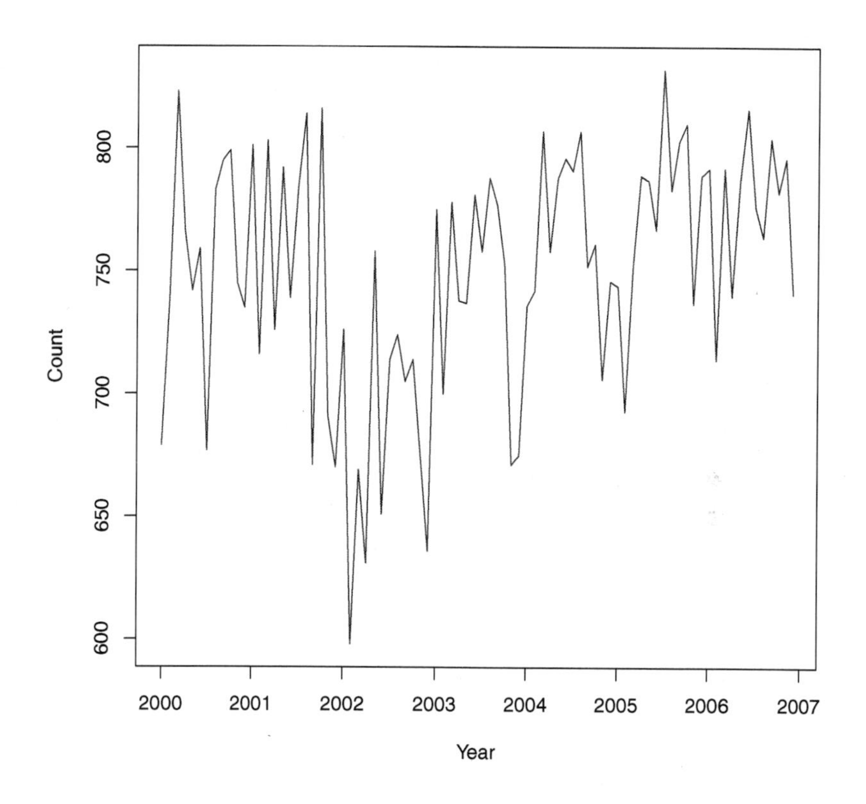

Figure 2.1: Monthly patient count for product related to medical problems.

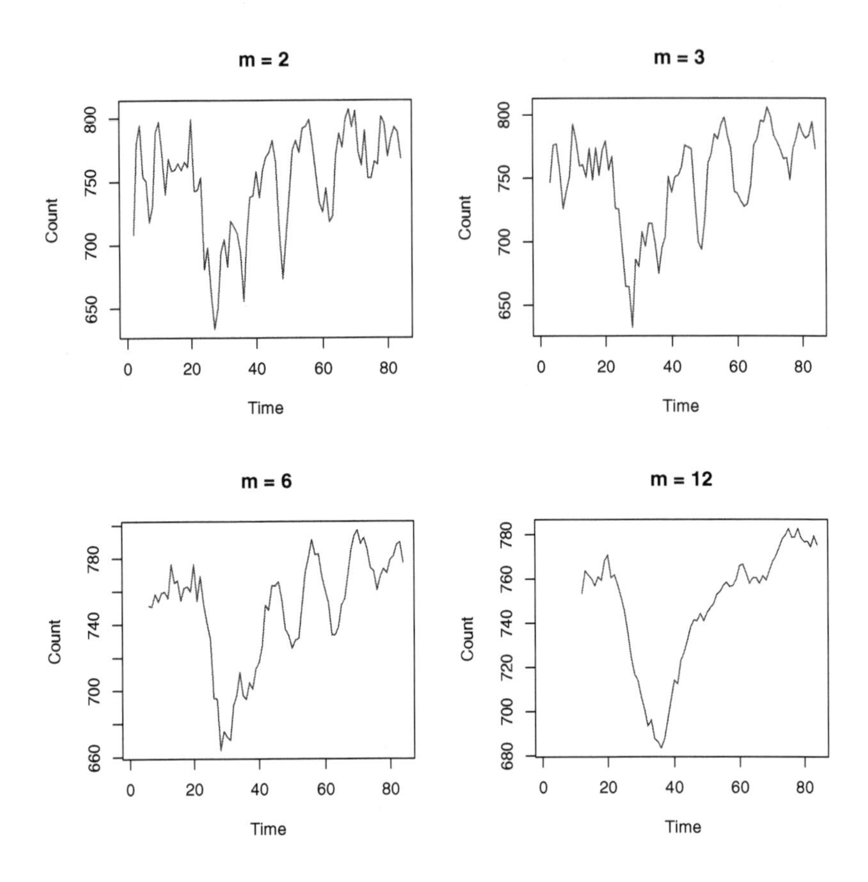

Figure 2.2: Series smoothed for different values of m

NOTE... ✍

The larger the order m, the greater the smoothing effect. If m becomes very large, the forecast collapses to the average value of the observed data.

Assessing Forecast Performance

How can we assess the performance of a forecasting model? In practice, there are a very large number of metrics to choose from. In this section, we discuss two popular metrics - the mean absolute error, and the mean absolute percent error.

Mean Absolute Error

The Mean Absolute Error (MAE) is the average of the absolute errors. The absolute error is the absolute value of the difference between the forecast value and the actual observed value:

$$MAE = \frac{1}{n}\sum_{i=1}^{n}\left|\frac{y_i - \hat{y}_i}{y_i}\right|$$

MAE tells us how large an error we can expect from the forecast on average.

Mean Absolute Percent Error

The Mean Absolute Percentage Error (MAPE) measures predictive accuracy as a percentage, and is defined by the formula:

$$MAPE = \frac{100}{n}\sum_{i=1}^{n}\left|\frac{y_i - \hat{y}_i}{y_i}\right|$$

MAPE is popular because of its very intuitive interpretation in terms of relative error. Higher values indicate lower predictive performance, and low values indicate higher predictive performance.

MAPE is scale-independent, and frequently used to compare forecast performance between different data sets. For example, a financial risk manager could compare the accuracy of a forecast of the FTSE 100 with a forecast of the S&P 500, even though these indexes may be at very different levels.

NOTE...

Business forecasters often report $100 - MAPE$ as an accuracy measure.

Advantages of Simple Moving Averages

The SMA model is often a good choice if the average of your time series observations are constant (stationary). This is because it is based on the assumption of a constant mean. For this reason, it can work well where your data fluctuate mildly around a constant level. In this case, the largest value of m will give the best estimates of future values. A longer observation period will average out the effects of variability. It can also work well when the data change very slowly over time.

Example - Predicting Medical Product Use

In the wonderfully crafted article "The Drug Pushers"published in *The Atlantic*, Carl Elliot recounts a memory of pharmaceutical sales representatives:

> "*Back in the old days, long before drug companies started making headlines in the business pages, doctors were routinely called upon by company representatives known as "detail men." To "detail" a doctor is to give that doctor information about a company's new drugs, with the aim of persuading the doctor to prescribe them.*"

Medical product sales are part of a massive industry, hospital, sales representatives and pharmaceutical companies are interested in forecasts of patient product use. In this section, we apply a simple moving average model to this task.

Step 1 – Collect, Explore and Prepare the Data

The hospital dataframe in the expsmooth package contains monthly patient counts for 767 products that are related to medical problems. First, load the dataframe and the packages:

```
data("hospital",package ="expsmooth")
```

Figure 2.1 shows the monthly patient count for the product in column 555. You can recreate the image using the following R code:

```
plot(hospital[,555],
xlab="Year",
col="darkblue",
ylab="Count")
```

Using decompose

We can see from the plot that this time series can (probably) be described using an additive model. This is because the random fluctuations in the data are roughly constant in size over time. The decompose function can provide additional insight:

```
decomp<-decompose(hospital[,555])
```

The R object decomp contains details of the decomposed series. Let's use the plot function to examine decomp visually:

```
plot(decomp)
```

You should see Figure 2.3. We can see this time series has some seasonal variation. The seasonal fluctuations are roughly constant in size over time, and do not seem to depend on the level of the time series. In addition, the random fluctuations are also roughly constant in size over time. This indicates the time series is well described using an additive model.

The random component appears to oscillate without any discernible pattern around the value zero. This is a (qualitative) indication of the appropriateness of using an additive decomposition.

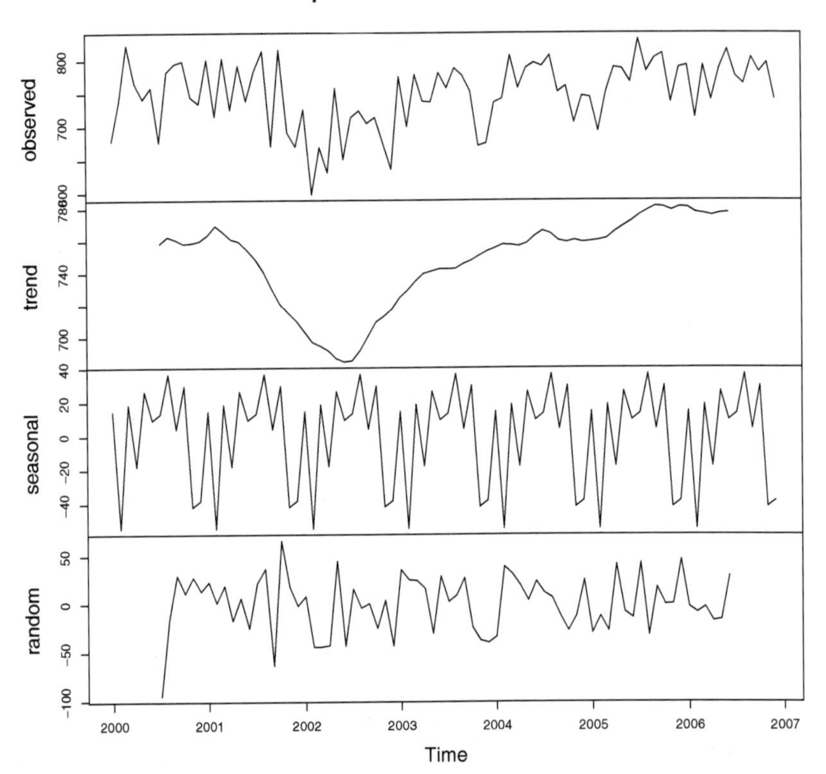

Figure 2.3: Decomposition of hospital product via `decompose`

Step 2 - Build a Forecast Model

The great thing about R is that it comes with many options for time series forecasting. We use the `sma` function from the `smooth` package to build our initial simple moving average model. The order, m, is the key parameter in `sma`. How should we choose it?

On the one hand, a larger value of m will reduce the effect of variability in the data due to the noise; On the other hand, a lower value of m will yield a forecast more responsive to changes in the mean level.

> ***NOTE...*** ✍
>
> The `smooth` package contains a set of smoothing functions used for time series analysis and forecasting. It uses a single error state space model.

The initial model

For our initial model, we set $m = 12$. We choose this value partly to capture the year over year dynamics of the monthly data. The model is built using the `sma` function. In a moment will type in the R code, but for now let's just walk through some features of the function. It takes a number of arguments:

- The first argument receives the sample data set. In this example, it is contained in the R object `data_sample`.
- The next argument is the order of the moving average, we set $m = 12$ via the `order` argument.
- Statistical prediction intervals are often useful in an initial analysis. We use a parametric interval set at a

95% confidence level. Other options for the parameter `interval` include - "`none`", "`nonparametric`" and "`semiparametric`"

- It is important to evaluate forecast accuracy by considering how well a model performs on data that were not used when fitting the model. It is common to use a portion of the available data for fitting – the in-sample data, and use the rest of the data to measure how well the model is likely to forecast on new data – the out-of-sample or hold out data. The `sma` function allows you to hold back some data for out of sample testing. This is achieved via the `holdout=TRUE` argument. The number of examples or sample points to hold out is controlled by the parameter `h`. We set it equal to 12, to set aside the most recent 12 months of data.

Now here is the R code:

```
data_sample<-hospital[,555]
require(smooth)
fit12<-sma(data_sample,
order=12,
h=12,
holdout=TRUE,
intervals="parametric",
level=0.95)
```

The model is fitted to the data very quickly, and you should see Figure 2.4. It presents the fitted SMA model, the actual data, and the predicted values. The fitted values are rather smooth, and capture the longer term trend in the data. However, they miss much of the month by month undulation. This is a fairly typical pattern, and is a direct consequence of setting the order to 12. As we saw earlier, the higher the order the smoother the moving average model will be.

The forecast period is denoted by the solid (red) vertical line. The predicted values initially rise and then begin to decline. The chart also includes the prediction interval. For this illustration, the observed values all lie inside the 95% prediction interval. However, the interval itself is rather wide which indicates a high degree of uncertainty in the predicted values.

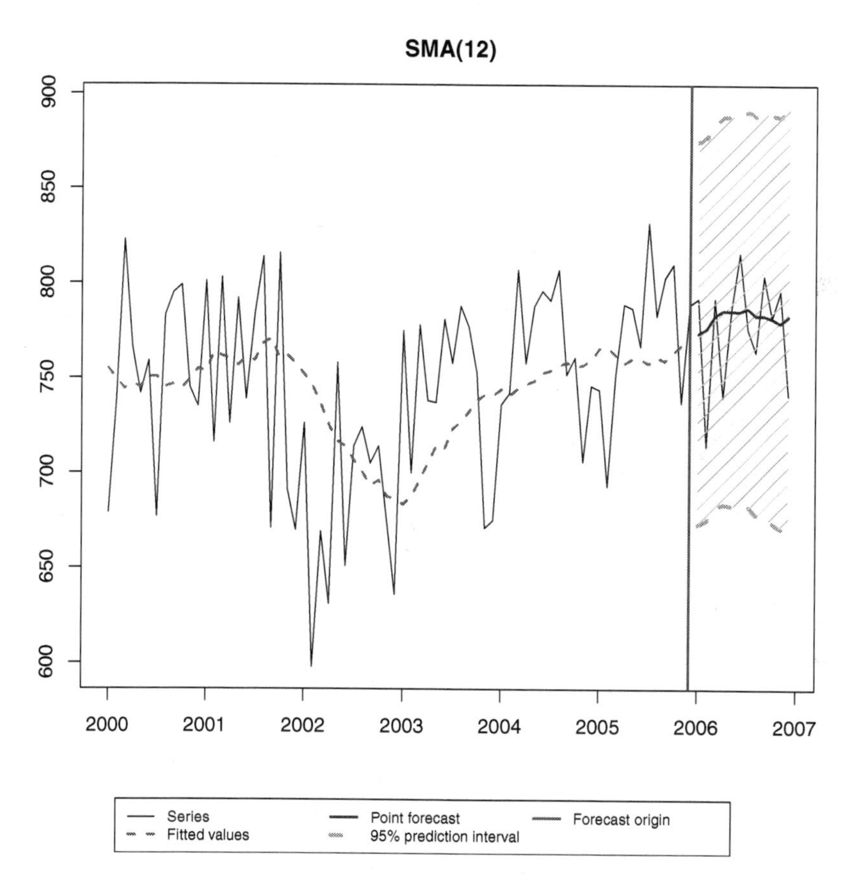

Figure 2.4: Fitted model and predicted values for `fit12`

NOTE... ✍

The `smooth` package builds forecasting models using a single source of error state-space model. The general idea behind any state-space model is that the process generating the time series observations consists of two parts. The first is the measurable or observable part. The second is an unobservable part which characterizes the evolution of observable components.

Step 3 - Evaluate Model Performance

The fitted values, predictions and other metrics are stored in the R object `fit12`. The `attributes` function provides a summary of its content:

```
attributes(fit12)

$names
 [1] "model"          "timeElapsed"
 [3] "states"         "transition"
 [5] "persistence"    "order"
 [7] "initialType"    "nParam"
 [9] "fitted"         "forecast"
[11] "lower"          "upper"
[13] "residuals"      "errors"
[15] "s2"             "intervals"
[17] "level"          "cumulative"
[19] "actuals"        "holdout"
[21] "intermittent"   "ICs"
[23] "logLik"         "cf"
[25] "cfType"         "accuracy"
```

```
$class
[1] "smooth"
```

Each of these attributes contain data associated with the fitted model. For example, to see the details of the fitted model:

```
fit12$model
```

```
[1] "SMA(12)"
```

This informs us we have fitted an SMA of order 12.

The holdout data and the forecasts can be viewed using a similar technique:

```
fit12$holdout
```

```
     Jan Feb Mar Apr May Jun Jul Aug Sep
2006 792 714 792 740 787 816 776 764 804
     Oct Nov Dec
2006 782 796 741
```

```
round(fit12$forecast,0)
```

```
     Jan Feb Mar Apr May Jun Jul Aug Sep
2006 774 776 783 786 786 785 787 783 783
     Oct Nov Dec
2006 782 779 783
```

Often, in traditional statistical analysis the residual (or model error) is assumed to be generated from a normal distribution (bell-shaped Gaussian distribution). Figure 2.5 illustrates a density plot of the model residuals. This is the difference between the fitted and actual values. Notice it does not have the neat bell shape of the Gaussian distribution. This is the norm for empirical data.

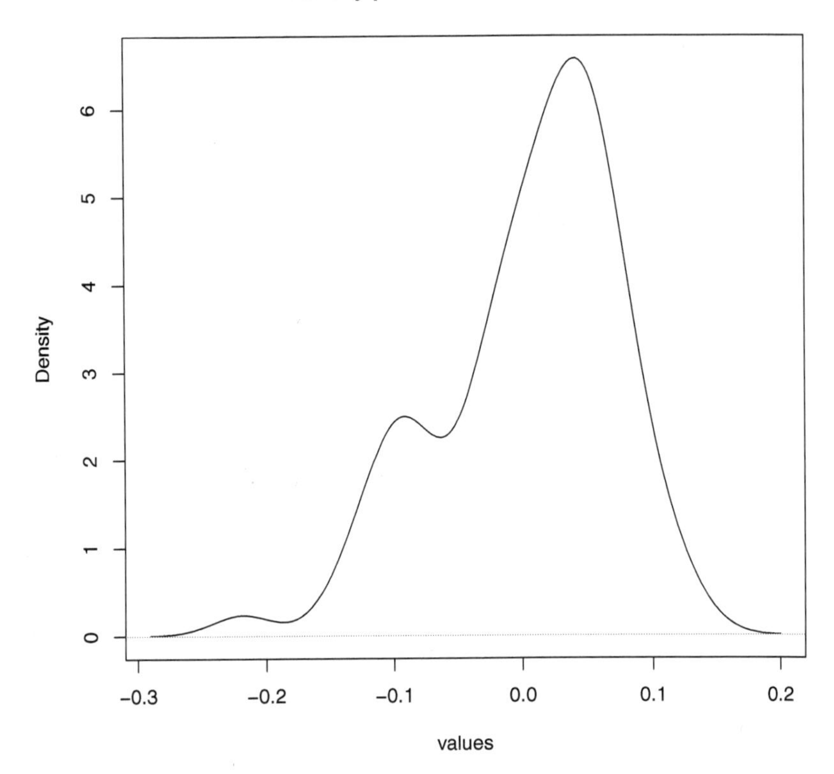

Figure 2.5: Density plot of residual for `fit12`

We assess model accuracy using MAE. We can use the `Metrics` package to perform the calculation for us:

```
require(Metrics)
round(mae(fit12$holdout,
fit12$forecast),2)
```

```
[1] 23.07
```

So, the MAE for `fit12` takes a value of 23.07. On its own this number has little value, however we can use it to benchmark alternative models. In other words, if we develop several models to forecast this series we would compare each of them to this value; and then select the model with the minimum MAE as our preferred model.

Step 4 - Improving Model Performance

Traditionally, the data analyst had to choose the order, m. If you have many models to build in a short period, manual selection of the order can be very cumbersome. For this reason, several simple rules of thumb were developed. For example, for monthly data one rule of thumb is to set $m = 12$. For quarterly data, you might build an initial model with $m = 4$.

Of course, these values are simply rules of thumb, and may not necessarily yield the optimal forecast. In practice, the process of choosing the order is iterative, and can take a considerable amount of time.

Fortunately, the `sma` function can choose the optimal order for you automatically using an information criterion (we discuss this topic more fully on page 73):

```
fit<-sma(data_sample,
h=12,
holdout=TRUE,
intervals="parametric",
level=0.95,
ic = "AIC")
```

The above R code is similar to that used for `fit12`. However, no order is specified, and the `ic` parameter is set to "AIC" in order to use Akaike information criterion for model selection.

The `ic` parameter

You can set the `ic` parameter to select your model selection criteria, for example set it to "`AIC`" for the Akaike information criterion (default), or "`AICc`" which corrects AIC for finite sample sizes. You can also set it to "BIC" to use the Bayesian information criterion.

The optimal model

Figure 2.6 indicates the optimal model has an order of 3. The fitted values closely mirror the dynamics of the underlying data. In this sense, the fit is much better than the first model.

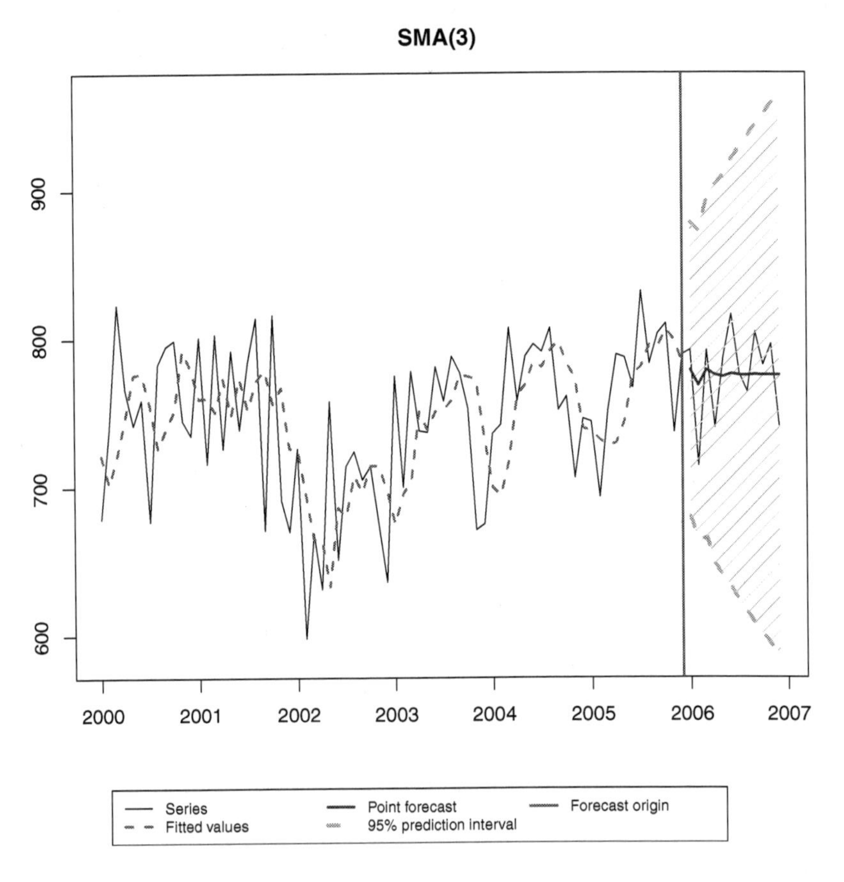

Figure 2.6: Automatic order selection for SMA

As with the previous model, we can use MAE to assess performance on the hold out sample:

```
round(mae(fit$holdout,
fit$forecast),2)
```

```
[1] 22.64
```

This is slightly smaller than the MAE of our first model. Of course, statistics like MAE cannot capture any qualitative interpretation of the performance of the model, nor does it capture concern about the large prediction intervals.

> ***NOTE...*** ✍
>
> Automatic model selection does not necessarily select the most usable model. You always have to assess the fitted model (automatic or manual) against intended use.

Trying an alternative order

It is nearly always worth trying alternative values. Let's build another model, this time with $m = 6$:

```
fit6<-sma(data_sample,
order=6,
h=12,
holdout=TRUE,
intervals="parametric",
level=0.95)
```

Figure 2.7 indicates the model has similar overall dynamics to the 3-month moving average model we developed in the R object `fit`. There is a touch more variation in the forecast, and the prediction intervals are not quite as wide. However, the model is unable to capture the wide swings in the month to month variation, instead it captures much of the trend.

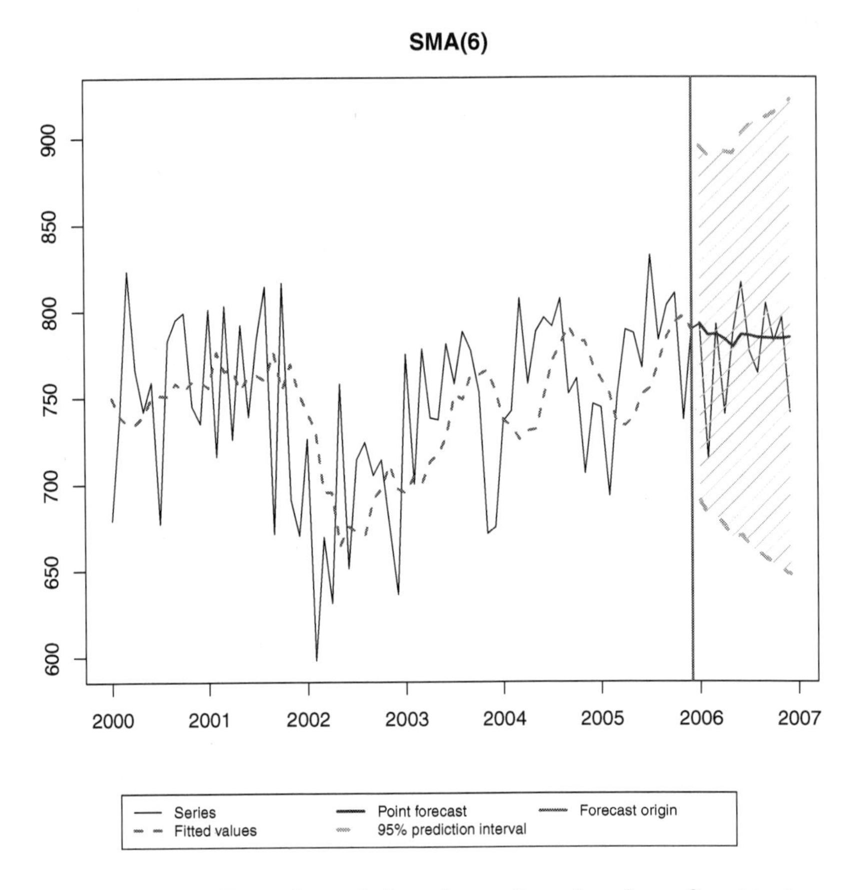

Figure 2.7: Fitted model and predicted values for `fit6`

As you can see below, the MAE is very similar to that of `fit`.

```
round(mae(fit6$holdout,
fit6$forecast),2)
```

```
[1] 22.21
```

Generally, you want to select the order that generates the least error between actual and predicted results. However, as in this example, there may be several candidates whose performance is similar. In the end, the choice between the three models may come down to a subjective opinion.

Limitations of Simple Moving Averages

Moving averages create a forecast by smoothing previous data using averages. The effects of seasonality, business cycles, and other irregular occurrences are not explicitly modeled. If seasonality is a significant proportion of the dynamics, simple moving average models can have large forecast errors. If this is the case you might consider using a model that caters for seasonality; we discuss many of these in the remainder of this text.

Summary

One of the easiest time series forecasting techniques to deploy is the simple moving average model. It is used to smooth out short-term fluctuations in a series of data in order to more easily recognize longer-term trends or cycles. Essentially, it estimates the next period's value by averaging the value of the previous m periods.

The technique is a simple, often very effective, tool for projecting future values of time series data. SMA models often perform well, and are easy to build and test using R.

In the next chapter, we look at another smoothing technique - exponential smoothing.

Chapter 3

Exploring Simple Exponential Smoothing

EXPONENTIAL smoothing is a very popular forecasting method. It was originally developed in the 1950s, and is used in retail, marketing, operations management and taught to in business schools. It is popular in part, because it can be easily programmed into a simple spreadsheet, and can generate highly accurate forecasts.

By the end of this chapter you will understand:

- How the simple exponential smoothing model works.
- Alternative methods such as Holt's method, and the Holt-Winters technique.
- Some metrics to assess model forecasting performance.
- How to build and automate exponential moving average models using R.

In general, exponential smoothing is an offshoot of the standard moving average technique we discussed in the previous chapter. It can be used to make short-term forecasts for time series data; and remains one of the fundamental time-series modeling techniques popular for its simplicity & low computational cost.

It is easy to understand and use, and most commercial forecasting software products include it in their offerings - including R.

Understanding Exponential Smoothing Methods

Earlier we saw the simple moving average model applied equal weight to each observation in its predictive equation. Exponential smoothing gives more weight to the most recent observations, and less weight to older values. The idea of weighting observations by time is popular in areas such as sales forecasting. This is because the most recent daily sales numbers may be more important than the numbers generated last quarter or the quarter before that.

Simple Exponential Smoothing

The simple exponential model (SEM) is a special type of weighted moving average often used to identify recent trends. It is given by:

$$\hat{y}_t = \alpha y_{t-1} + \alpha(1-\alpha) y_{t-2} + \alpha(1-\alpha)^3 y_{t-3} + ...$$

where α is called the smoothing constant, and takes values $0 < \alpha < 1$. From this equation, you can see that the method constructs a weighted average of the observations. Notice that the formula above is recursive. This means all the past values end up playing a role in the forecast, although the weight of each observation decreases exponentially as we move backwards in time. In other words, decreasing importance is given to values as they get "older" in time.

After a little algebra, we see that:

$$\hat{y}_t = \alpha y_{t-1} + (1-\alpha)\left[\alpha y_{t-2} + \alpha(1-\alpha) y_{t-1} + ...\right]$$

which can be further simplified to:

$$\hat{y}_t = \alpha y_{t-1} + (1-\alpha)\,\hat{y}_{t-1} \tag{3.1}$$

From equation 3.1, we see that the forecast for the next time period is a weighted sum of the previous forecast ($\hat{y}_t$) and previous actual value (y_{t-1}) where the weights are determined by the smoothing parameter α.

NOTE...

The age of data in SEM is approximately $\frac{1}{\alpha}$.

An alternative view

We can rewrite 3.1 as:

$$\hat{y}_t = \hat{y}_{t-1} + \alpha\epsilon_{t-1} \tag{3.2}$$

where $\epsilon_{t-1} = (y_{t-1} - \hat{y}_{t-1})$ is a measure of the error between the observed and forecast value.

From equation 3.2 we see that each new forecast is simply the old forecast plus an adjustment for the information contained in the error $(y_{t-1} - \hat{y}_{t-1})$ from the last forecast. Small values of α cause the forecasts to react slowly to the error in past forecasts. Large values cause a much more rapid adjustment. This adjustment process is analogous to an electronic thermostat where corrections are made to alter the temperature if it is either too hot or too cold.

The forecast includes all past observations where the more recent observations are more heavily weighted than older observations. For example, the most recent observation y_{t-1} receives a weight of α. The third most recent observation y_{t-3} only receives a weight of $\alpha(1-\alpha)^3$.

In other words the smoothing constant, α, determines how fast the weights of the time series decay. This idea is illustrated in Figure 3.1.

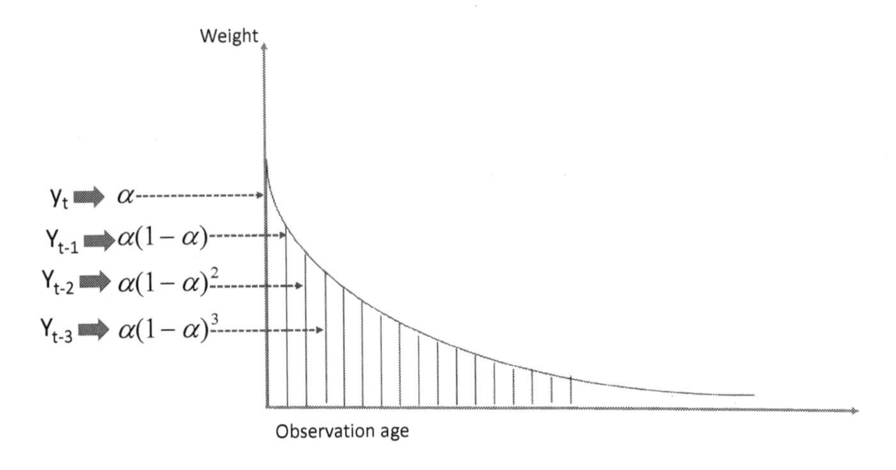

Figure 3.1: Simple Exponential Smoothing weights

> ***NOTE...***
>
> The term "*simple exponential smoothing model*" came from the observation that averaging is a form of smoothing and the weights decrease exponentially through time.

A simple illustration

As an illustration, suppose $\alpha = 0.5$ then y_{t-1} has a weight of 0.5, and y_{t-3} has a weight of 0.0625. Notice that:

- The larger the value of α, the more responsive the forecast in the sense that $\hat{y}_t$ will more closely match y_{t-1}. Values of α close to 1 give greater weight to the most recent observed values.

- A smaller value of α, results in a smoother forecast. Values close to 0 allow distant past observations to have a larger influence. This makes the forecast less responsive to recent changes in the observed values.

Values of α that are close to 0 mean that little weight is placed on the most recent observations when making forecasts of future values. The value of α is usually found automatically by minimizing the fitted value forecast error. Alternatively, a popular rule of thumb sets it manually to lie between 0.1 and 0.3.

Relationship to SMA

A small smoothing constant α has a similar smoothing impact as selecting a large order m in the SMA. Both result in a smoothed series which is less responsive to changes in present and future values. Conversely, a large α has similar properties to a small order in the SMA.

It turns out that if we set $\alpha = \frac{2}{m+1}$, then the SMA and SEM will have similar variances of the errors.

NOTE... ✍

You can think of the SEM forecast $\hat{y}_t$ as a weighted average of all previous observations. The weights decrease exponentially with the age of the data.

Assessing Forecast Performance

A simple measure of how well a model fits the observed data is known as the sum of squared errors:

$$SSE = \sum_{i=1}^{n} (y_i - \hat{y}_i)^2$$

It is a measure of how close the forecast values ($\hat{y}_i$) match the actual observations (y_i). The difference between $\hat{y}_i$ and y_i is the error in the forecast. Hence, the term sum of squared errors. The smaller the SSE, the more closely the predicted values match the actual observations.

Mean Squared Error

An alternative to the SSE is the Mean Square Error (MSE).

It is derived by summing the squared differences between the observed observation (y) and predicted value ($\hat{y}$), then dividing by the number of examples (n):

$$MSE = \frac{1}{n} \sum_{i=1}^{n} (\hat{y}_i - y_i)^2$$

A good way to think about MSE is that it measures the average ($\frac{1}{n} \sum_{i=1}^{n}$) of the squares of the errors between the observed target and predicted values ($(\hat{y}_i - y_i)^2$).

The smaller the MSE, the closer the predictions match the observed target values. It penalizes large errors relatively more than small errors because it squares them first.

The errors are squared so negative values do not cancel positive values. Of course, this also implies MSE is measured in squared units of the original observations.

> ***NOTE...*** ✍
>
> MSE is approximately equal to the standard deviation of the errors if the mean error is close to zero.

Root Mean Squared Error

The Root Mean Square Error (RMSE) (also called the root mean square deviation) is another frequently used measure of the difference between values observed and predicted values. Since the MSE is measured in squared units of the target, it is common practice to report the square root of this value, known as the root mean squared error:

$$RMSE = \sqrt{MSE}$$

RMSE is a little easier to interpret because it is measured in the same units as the original observations. It can be interpreted as the average distance, between the predicted and observed values, measured in units of the target variable.

> ***NOTE...*** ✍
>
> RMSE and MAE are frequently compared to determine whether the forecast contains large but infrequent errors. The larger the difference between RMSE and MAE the more inconsistent the error size.

Holt's Method

Holt's method is a simple technique that is often used in place of SEM when the data has a linear trend. It estimates smoothed versions of the level and the trend of the time series. These components are combined and extrapolated forward to obtain the forecast.

Level equation

There are three equations used to define this technique. The first is the exponentially smoothed series:

$$A_t = \alpha y_t + (1 - \alpha)(A_{t-1} + T_{t-1}) \tag{3.3}$$

where $0 < \alpha < 1$.

This is similar to the SEM model where α serves as the smoothing constant. However, an estimate of the trend (T) is also included, and the smoothed series generated from the observed values (y_t) is captured by A_t.

Trend Equation

The trend is estimated by taking the difference between consecutive values of the smoothed series, i.e. by taking the difference of $A_t - A_{t-1}$ using the equation:

$$T_t = \beta\,(A_t - A_{t-1}) + (1 - \beta)\,T_{t-1} \tag{3.4}$$

where $0 < \beta < 1$.

This is similar to equation 3.3 except that the smoothing is applied to the trend. The smaller the value of β, the larger the weight given to the previous trend estimate T_{t-1}.

NOTE... ✍

Holt's method is often referred to as a "double" exponential smoothing because it estimates both the level (A_t) and the trend (T_t).

Forecast Equation

Finally, equation 3.3 and equation 3.4 are combined into the predictive equation:

$$\hat{y}_{t+h} = A_t + hT_t \tag{3.5}$$

The constant h is the number of periods to forecast into the future.

Holt's method is similar to SEM except that it smooths the trend and the slope using different smoothing constants.

Low values of α and β should be selected if there are frequent random fluctuations in the data and little or no trend.

NOTE... ✍

The value of α and β are often determined by minimizing a measure of error, usually the Mean Square Error (MSE).

Holt-Winters Method

The Holt-Winters (HW) method is one of the most flexible tools for modeling using exponential weights. It is a fundamental approach used in business forecasting.

To produce a forecast, HW estimates three components.
The first is the smoothed time series using a similar approach to that we observed earlier with the SEM and Holt's method.

Second, the trend is estimated as the change in the underlying level that we expect between today (t) and the next period ($t + 1$). For example, if we estimate our current level at 20 units, and we expect it to be 30 units for the next period, then our estimated trend is +10 units.

Finally, HW adjusts for seasonality (which neither SEM nor Holt's method were capable of capturing).

As we saw on page 11, a time series can be decomposed into multiplicative or additive. The HW method captures this dynamic by using different equations depending on whether the data is multiplicative or additive.

NOTE... ✍

The HW method is also referred to as exponential smoothing with additive trend and seasonal component.

Additive Model

The additive HW model consists of four basic equations. For the additive model. The exponentially smoothed series is obtained via:

$$A_t = \alpha\left(y_t - s_{t-c}\right) + (1-\alpha)\left(A_{t-1} + T_{t-1}\right) \tag{3.6}$$

where s_{t-c} is the estimate of the seasonal factor at time $t-c$, with c capturing the number of periods typically in a year. For example, $c = 12$ for monthly data, and $c = 4$ for quarterly data.

The equation updates the smoothed series by taking the difference between the observed y_t value at t and the seasonal s_{t-c} factor. This adjusts the data for seasonality.

The trend component is estimated using equation 3.4; and the seasonality estimate is given by:

$$s_t = \gamma\left(y_t - A_t\right) + (1-\gamma)\, s_{t-c} \tag{3.7}$$

After the trend and seasonal components have been smoothed. The forecast for h periods ahead is calculated:

$$\hat{y}_{t+h} = A_t + hT_t + s_{t-c+h} \tag{3.8}$$

For each component (level A_t, trend T_t, seasonal s_{t-c}) there is a smoothing parameter (α, β and γ), that takes values between 0 and 1.

Similar to the SEM model we saw earlier, smaller values of these smoothing constants place more weight on the smoothed values and less on the observed value. Larger smoothing parameter weights allow the method to adapt more quickly to short term changes in the time series pattern, but there is a risk the forecast overreacts to unusual values.

NOTE... ✍

The smoothing parameters α, β and γ are usually estimated by minimizing the MSE over the observations.

Multiplicative Model

The exponentially smoothed series for the multiplicative HW model is obtained via:

$$A_t = \alpha \left(\frac{y_t}{s_{t-c}} \right) + (1 - \alpha) (A_{t-1} + T_{t-1}) \tag{3.9}$$

where s_{t-c} is the estimate of the seasonal factor. The trend component is estimated using equation 3.4; and the seasonality estimate by:

$$s_t = \gamma \left(\frac{y_t}{T_t} \right) + (1 - \gamma) s_{t-c} \tag{3.10}$$

After the trend and seasonal components have been smoothed. The forecast for h periods ahead is calculated:

$$\hat{y}_{t+h} = (A_t + hT_t) s_{t-c+h} \tag{3.11}$$

As a simple illustration, suppose we are interested in a forecast for the next period (h = 1), and that the smoothed level $A_t = 1000$, with trend $T_t = 10$, and seasonal factor s estimated to take the value 1.2. The forecast for the next period is simply:

$$(1000 + 10) \times 1.2 = 12,000$$

Advantages of Exponential Smoothing

The exponential smoothing models are easy to understand and useful for short-term forecasts. They give more weight to recent observations, and diminish the weight on distant observations. This characteristic is especially useful in forecasting business time series such as sales data.

Since SEM uses all the data, it does not experience jumps in the forecast when observations fall out of a window (as can be the case when using a simple moving average model).

Example - Modeling Industrial Production

The breathtaking coast line, lush valleys and majestic mountains make Cantabria a wonderful place to visit. It is located on Spain's north coast, and a hub of economic activity. In this section we develop a number of exponential smoothing models to forecast industrial production in this autonomous Spanish region.

Step 1 – Collect, Explore and Prepare the Data

The data we use is stored in the R object `ipi` from the package `descomponer`. First, load the data:

```
data("ipi",package ="descomponer")
```

Figure 3.2 plots the index of industrial production (monthly) over the period 2001 to 2014. The data contain seasonal patterns, and the level appears to rise slightly from 2002 reaching a peak around 2008, before declining. Overall the series is volatile, reflecting the inherent uncertainty associated with economic activity.

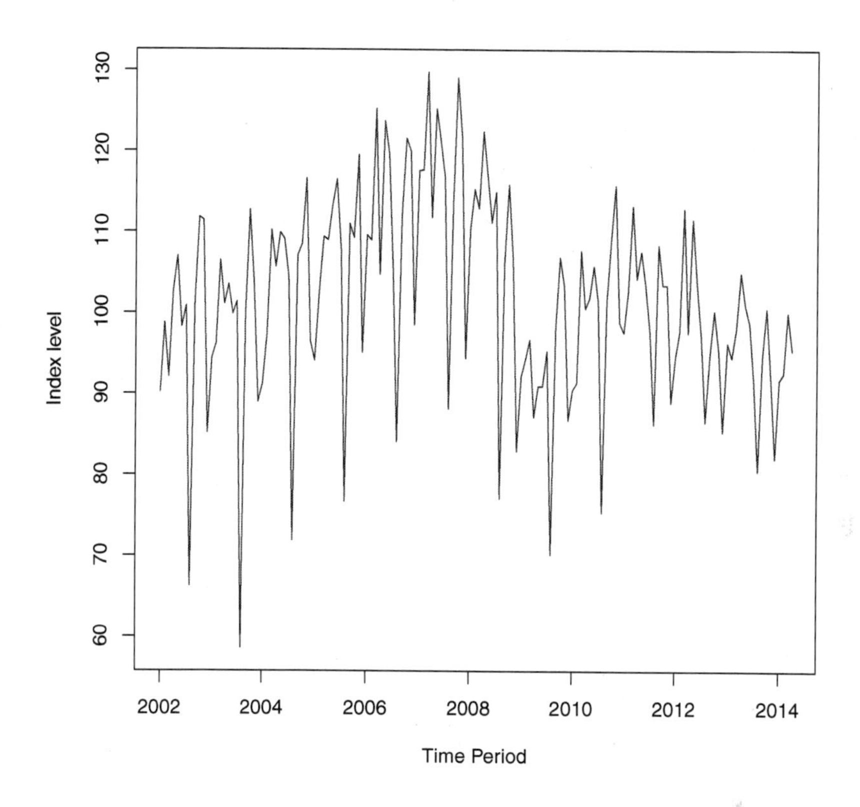

Figure 3.2: Index of Cantabria industrial production

It appears evident the sample data may be well described using an additive model. Let's try that specification with the decompose function, to obtain a visual representation of the series components:

```
data_sample<-ts(ipi,
frequency=12,
start = c(2002, 1))
plot(decompose(data_sample))
```

The first statement stores the ipi data as time series in the R object data_sample. The statement line simply calls the plot function.

You should see Figure 3.3. The random fluctuations in the time series seem constant, so it is probably appropriate to describe the data using an additive model. The image also confirms a modest hump shaped trend, along with seasonality.

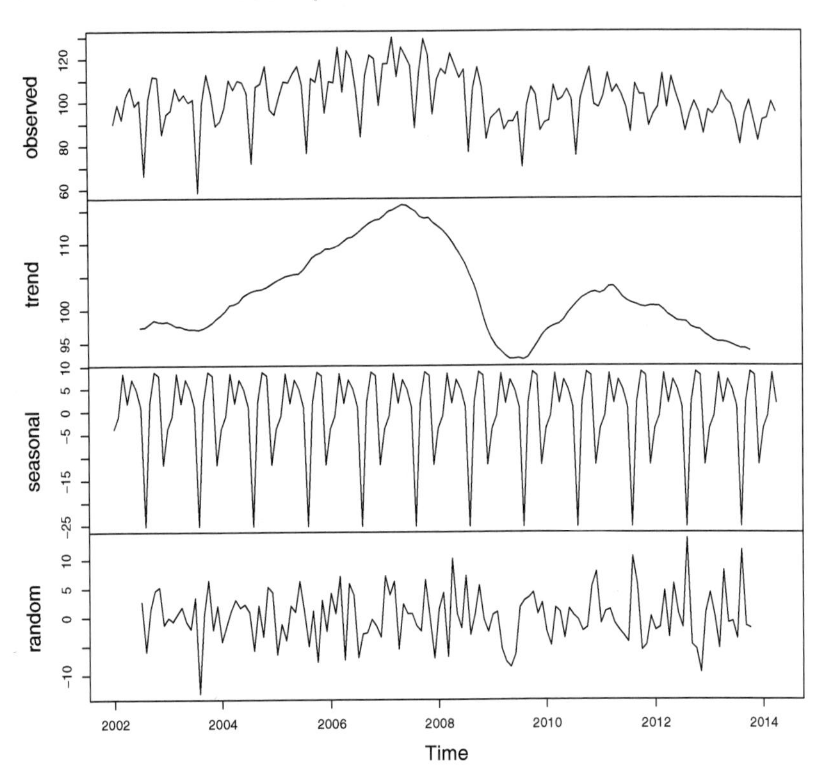

Figure 3.3: Use of `deccompose` with `ipi` data

Step 2 - Build a Forecast Model

We build a model to forecast the final 12 months of the data sample. Our first step, is to create the training set sample. This data is stored in the object `data_train`:

```
n=length(ipi)
n_test=12
n_train=n-n_test
data_train<-ts(ipi[1:n_train],
frequency=12,
start = c(2002, 1))
```

The training data is stored as a time series object via the `ts` argument.

Building a Simple Exponential Model

The `HoltWinters` function comes with the basic installation of R (`stats` package). It can be used to fit exponential smoothing models.

It takes the sample data and three parameters - `alpha`, `beta` and `gamma`. A simple exponential model is fitted by setting both `beta` and `gamma` to "`FALSE`", and then specifying a value for `alpha`.

Let's try a simple exponential model, with `alpha = 0.3`. Here is how to specify it:

```
fit_es1<-HoltWinters(data_train,
alpha=0.3,
beta=FALSE,
gamma=FALSE)
```

Figure 3.4 shows the fitted and actual values. The fitted values follow the general trend in the data, but (as expected) miss the extreme movements driven by seasonality.

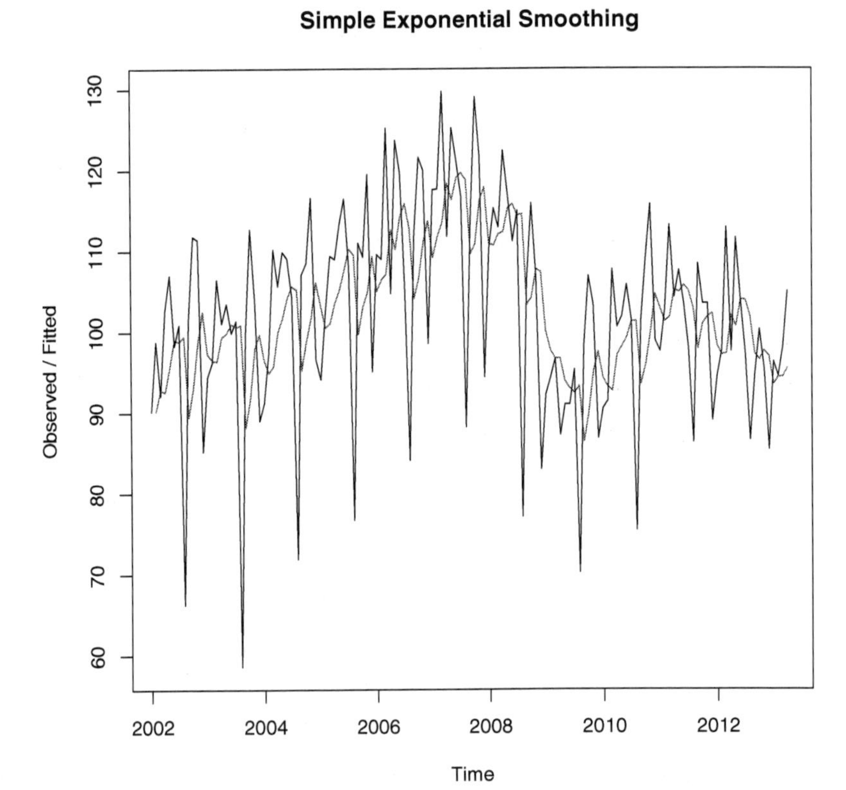

Figure 3.4: `fit_es1` fitted and actual values

Sum of squared errors

A nice feature of the `HoltWinters` function is that it also reports the sum of squared errors for the fitted model. A simple way to see this value is to use:

```
fit_es1$SSE
```

```
[1] 20436.75
```

What does this value tell us? Not very much on its own. It is best used as a benchmark against which to compare alternative models. The smaller the value, the better fitting the model.

Automatic estimation of the smoothing parameter

In the traditional approach to building exponential smoothing models, the analyst manually selects the smoothing parameter. We did this in the above example. However, when you have a lot of time series to forecast this approach can be very inefficient.

Fortunately, the smoothing parameter can be determined automatically. The `HoltWinters` function chooses the value of α by minimizing the squared one-step prediction error. Simply, pass the data to the function, but drop the `alpha` argument as follows:

```
fit_es<-HoltWinters(data_train,
beta=FALSE,
gamma=FALSE)
```

To view the estimated value of α append `$alpha` to `fit_es`:

```
round(fit_es$alpha,4)
```

```
[1] 0.1454
```

The output informs us that the estimated value of the α parameter is about 0.15. This tells us that relatively less weight is placed on recent observations than our initial guess of 0.3. Automatic model selection is useful therefore, to assess our own intuition and determine feasible values for the smoothing parameter.

Figure 3.5 shows the fitted and actual values. The fitted values are much "smoother" than those in Figure 3.4. This is a direct consequence of placing relative less weight on the most recent observations.

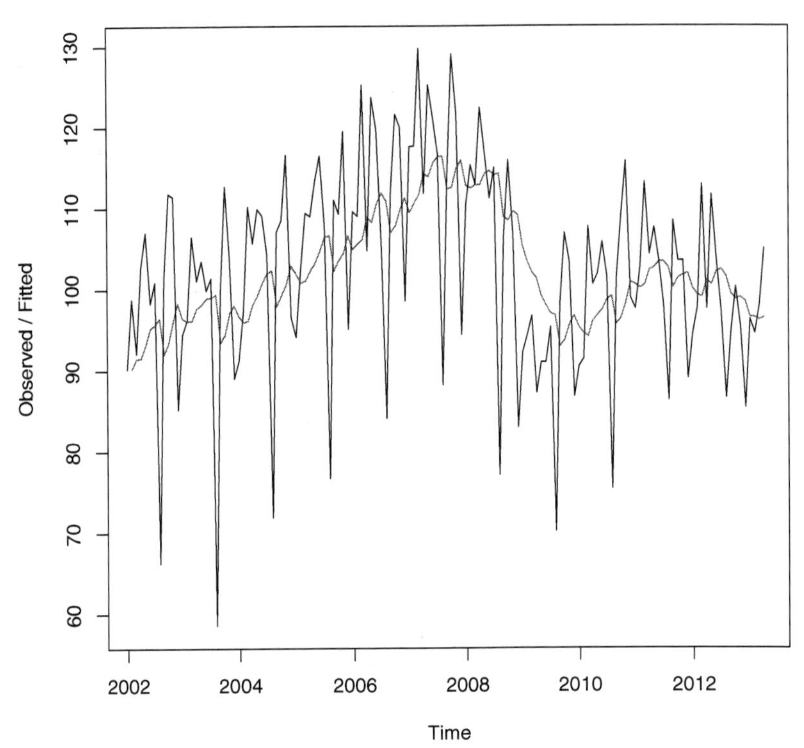

Figure 3.5: `fit_es` fitted and actual values

Step 3 - Evaluate Model Performance

If you compare Figure 3.4 with Figure 3.5, it is difficult to tell which model has the better fit. As expected Figure 3.5 is smoother (see equation 3.1). However, to the untrained eye, both are quite similar as they capture the trend dynamics, but miss the seasonality component.

This is where the sum of squared errors for the fitted model can be useful. Earlier we saw the value for `fit_es1`. Here is the value for `fit_es`:

```
fit_es$SSE
```

```
[1] 19301.17
```

It measures in-sample degree of fit for the model `fit_es`. Since the value is smaller than that of `fit_es1`, we tentatively conclude that `fit_es$SSE` represents an improvement.

Most of the time you will want to choose the smoothing parameters automatically. Even if you intend to fine tune their values manually, automatic selection usually gives you a great place from which to start tweaking.

Trying Holt's Method

Since the data exhibit a mild trend, we could also use Holt's method. Here is how to fit it with the smoothing parameters determined automatically via the `HoltWinters` function:

```
fit_he<-HoltWinters(data_train,
gamma=FALSE)
```

The exponentially smoothed series parameter alpha, and the trend smoothing parameter beta are estimated as:

```
round(fit_he$alpha,4)
```

```
 alpha
0.2221
```

```
round(fit_he$beta,4)
```

```
  beta
0.3096
```

The estimated value of alpha is 0.22, and of beta is 0.31. These values tell us that both the estimate of the current value of the level (α), and of the slope of the trend component (β), are based less on very recent observations in the time series and more on the smoothed value. This makes intuitive sense, since the level and the trend in the time series change moderately over time.

The fitted and actual values are illustrated in Figure 3.6. You can see that the fit appears particularly poor throughout 2002 to 2004. It considerably overestimates the actual values throughout 2002. Overall, visually it appears to perform much worse than either of the two models we have already developed.

Let's take a look at the sum of squared errors:

```
fit_he$SSE
```

```
[1] 24114.34
```

Yep, this is considerably higher than either of our previous models.

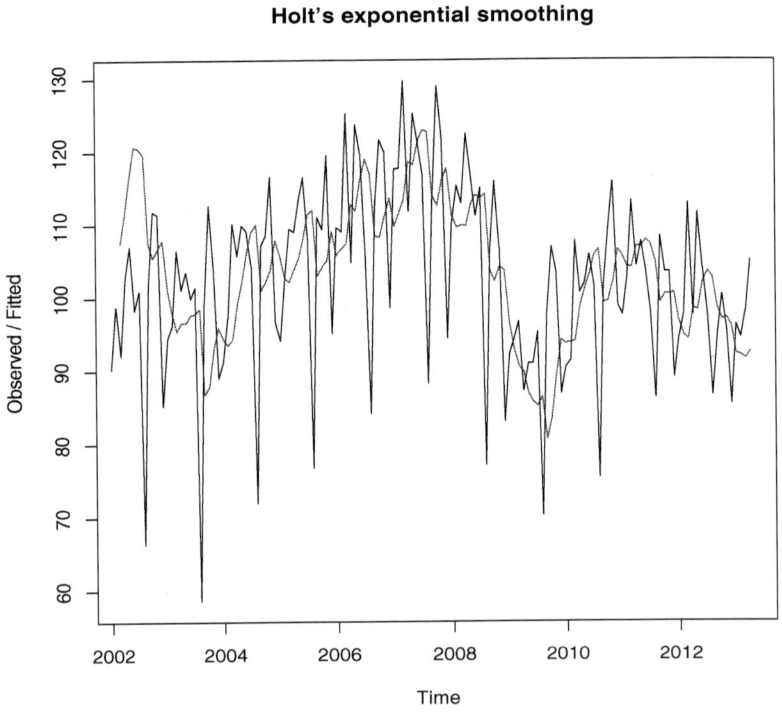

Holt's exponential smoothing

Figure 3.6: Holt's method fitted and actual values

Step 4 - Improving Model Performance

The chief difficulty with the models we've applied so far is their inherent inability to capture seasonality. Sense seasonality is a major characteristic of the observed time series we should try a model that can cater for it. Given these characteristics, we might expect the additive Holt-Winters model to perform much better than the simple exponential model or Holt's method.

Fortunately, the Holt-Winters model can be fitted with one line of R code:

```
fit_hw<-HoltWinters(data_train)
```

The model is trained in moments and the estimated parameters (α, β and γ) can be viewed directly.

First, the estimated value of α:

```
round(fit_hw$alpha,4)
```

```
0.327
```

The estimated value of α at 0.327 is relatively low, indicating that the estimate of the level at the current time point is based upon both recent observations and some observations in the more distant past.

Now for β:

```
round(fit_hw$beta,4)
```

```
0
```

The value of β is 0.00, indicating that the trend component is set equal to its initial value. This makes intuitive sense, as the trend component changes slowly over the time period.

Finally, lets take a look at γ:

```
round(fit_hw$gamma,4)
```

```
0.3356
```

The value of γ at 0.3356 indicates a strong role for previous seasonal component values relative to the recent difference between the observed value y_t and smoothed value A_t.

Sum of square and model fit

Of course, our interest in developing this model is to enhance performance by making use of the seasonality component. We can view the in-sample sum of squared errors (as we did earlier) as follows:

```
round(fit_hw$SSE,2)
```

```
[1] 4335
```

Wow! A massive reduction relative to the earlier models. Figure 3.7 shows the predicted and fitted values. From the plot, it is evident that:

- The Holt-Winters exponential method is very successful in predicting the seasonal peaks and troughs.
- The Holt-Winters model is much better at capturing the dynamics of industrial production than the simple exponential smoothing model or Holt's method.

This reminds us of the importance of using a forecasting model that can capture seasonality if it is evident in the data.

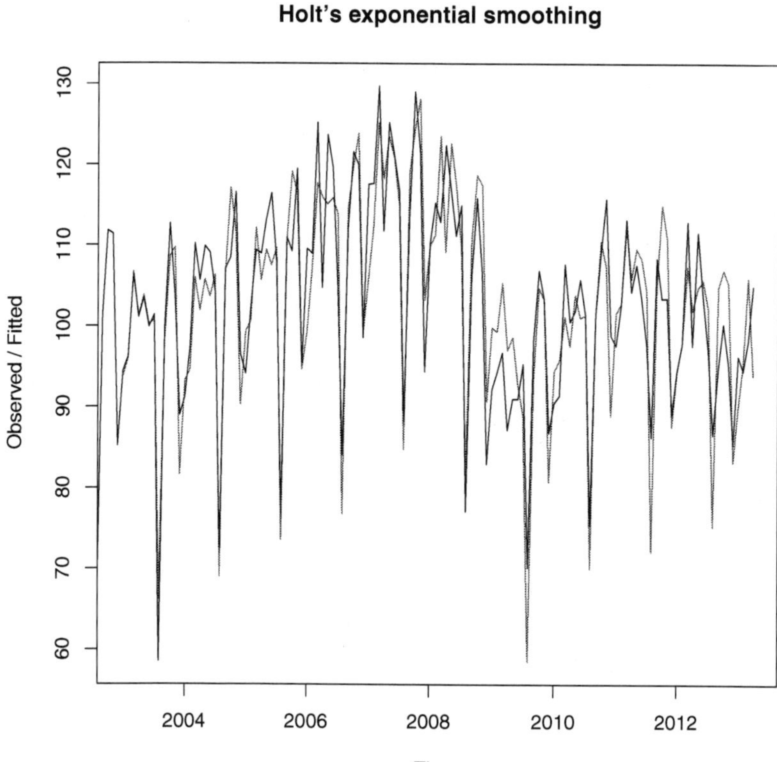

Figure 3.7: Holt-Winters method fitted and actual values

Assessing predictive performance

Next, we assess how well each of the models performed on the test sample. The `predict` function is useful here. For example, for Holt's model we can specify:

```
pred_he <- predict(fit_he,
n.ahead = 12,
prediction.interval = TRUE,
level = 0.95)
```

The above code passes the fitted model `fit_he` to the `predict` function, and sets the number of forecasts ahead (**n.ahead** parameter). It is often useful to include statistical prediction intervals, you can do this by toggling the parameter **prediction.interval**. The confidence level is set to 0.95.

The object **pred_he** contains the predicted values alongside the prediction interval. The first few values can be viewed using the **head** function:

```
round(head(pred_he),2)
```

```
           fit    upr   lwr
May 2013 95.64 121.97 69.31
Jun 2013 95.97 123.40 68.55
Jul 2013 96.30 125.32 67.29
Aug 2013 96.63 127.76 65.51
Sep 2013 96.97 130.73 63.20
Oct 2013 97.30 134.21 60.39
```

Let's repeat the process for the simple exponential model and the Holt-Winters model:

```
pred_es <- predict(fit_es,
n.ahead = 12,
prediction.interval = TRUE,
level = 0.95)

pred_hw <- predict(fit_hw,
n.ahead = 12,
prediction.interval = TRUE,
level = 0.95)
```

Earlier we mentioned that the RMSE is often used to assess predictive performance. We can calculate its value via the **Metrics** package. For example, for the Holt's model predictions contained in **pred_he**:

```
data_test<-ts(ipi[(n_train+1):n],
frequency=12, end = c(2014, 4))
require(Metrics)
round(rmse(pred_he[,"fit"],data_test),2)
```

```
[1] 7.73
```

Similarly, for the simple exponential model we have:

```
round(rmse(pred_es[,"fit"],data_test),2)
```

```
[1] 7.74
```

As we might expect both simple exponential smoothing model and Holt's method have almost identical root mean square. This confirms our earlier findings that the predicted performance between these two models is very similar.

Now, lets take a look at the RMSE for the Holt-Winters model:

```
round(rmse(pred_hw[,"fit"],data_test),2)
```

```
[1] 4.83
```

Since smaller is better, the Holt-Winters model is the clear winner. Seasonality is an important characteristic of the data and the Holt Winters model does a great job in capturing it relative to the other models we have discussed so far.

As can be seen in Figure 3.8 the forecasts match closely the actual observed values. It is also pleasing to notice the realized observations all lie within the 95% prediction interval. This provides additional statistical support for the model.

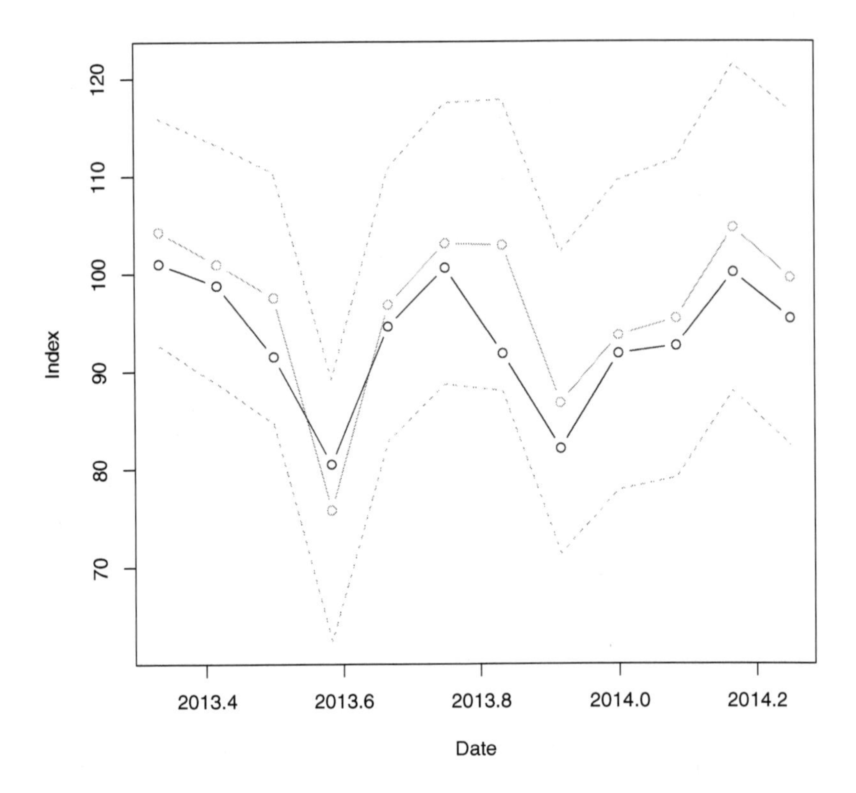

Figure 3.8: Forecast and actual values using the Holt-Winters method

Out of sample forecasts

Much of the time your forecasts will need to extend beyond the available data. One way to achieve this is to use the `forecast.HoltWinters` from the `forecast` package.

To illustrate, we refit a Holt-Winters model using all the industrial production data:

```
data_sample<-ts(ipi,
frequency=12, start = c(2002, 1))
```

```
fit<-HoltWinters(data_sample)
```

Notice, we use the R object data_sample to capture the entire data sample contained in ipi. The fitted model is stored in the R object fit.

Next, call the forecast.HoltWinters function:

```
require(forecast)
fit_outSample<- forecast.HoltWinters(fit_hw
,h=12)
```

In this example, we generate out of sample forecasts for the next 12 months (hence h=12).

To view the forecast values, alongside their prediction intervals use the print function:

```
round(print(fit_outSample),1)
```

	Point Forecast	Lo 80	Hi 80	Lo 95	Hi 95
May 2013	104.3	96.6	111.9	92.6	115.9
Jun 2013	101.0	93.0	109.0	88.8	113.2
Jul 2013	97.6	89.2	106.0	84.8	110.4
Aug 2013	75.8	67.1	84.6	62.4	89.2
Sep 2013	96.9	87.8	105.9	82.9	110.8
Oct 2013	103.1	93.7	112.5	88.7	117.5
Nov 2013	102.9	93.1	112.6	88.0	117.8
Dec 2013	86.8	76.7	96.8	71.4	102.1
Jan 2014	93.7	83.3	104.1	77.8	109.5
Feb 2014	95.4	84.7	106.1	79.1	111.7
Mar 2014	104.7	93.7	115.6	87.9	121.4
Apr 2014	99.5	88.3	110.8	82.4	116.7

The output reports the 80% and 95% level prediction interval as well as the forecast values (Point Forecast). Figure 3.9 represents this data visually.

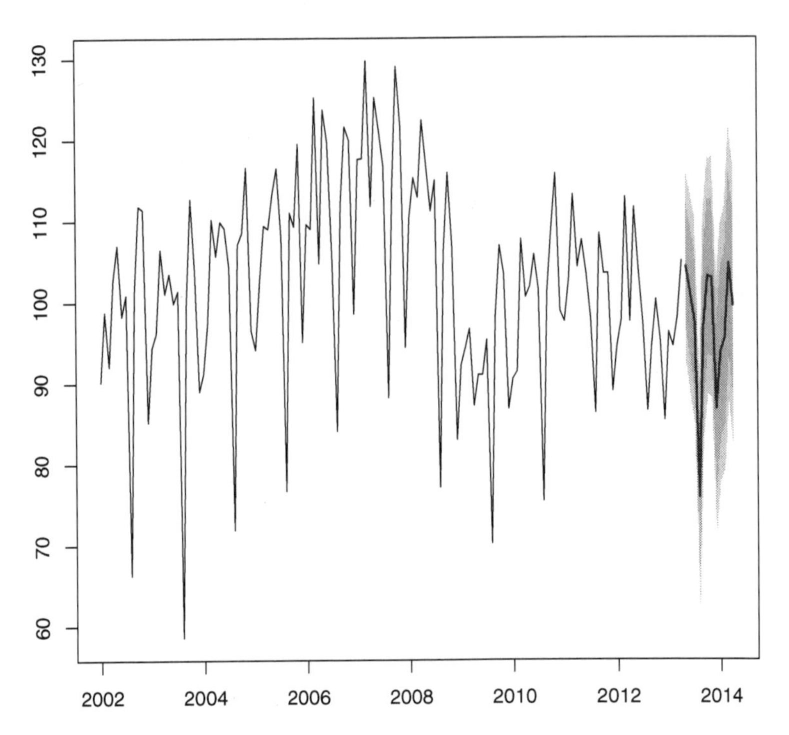

Figure 3.9: Out of sample forecast for Cantabria industrial production using Holt-Winters

Limitations Exponential Smoothing

The SEM model can perform poorly when there is a strong trend or seasonality. However, the use of Holt's method or the HW method can alleviate this issue.

Similar to SMA, SEM forecasts will lag behind the observed values especially if the data begin to develop a sharp trend upwards or downwards.

Rules of thumb

Several simple rules of thumb can help you determine between SEM, Holt's method and HW. If you have a time series that can be described using an additive model and you observe:

- no trend or seasonality, you can use simple exponential smoothing to make short-term forecasts.
- increasing or decreasing trend with no evidence of seasonality, you can use Holt's exponential smoothing to make short-term forecasts.
- increasing or decreasing trend and seasonality, you can use Holt-Winters exponential smoothing to make short-term forecasts.

Summary

Exponential smoothing type models have been around since the 1950s, and they continue to dominate business forecasting to this day. They are straightforward to use, and the forecasting procedure can take place automatically. This is especially useful in situations where you have numerous individual time-series for which forecasts are required. For example, in inventory control automated forecasting procedures are the most efficient option.

As you've seen, the models are very easy to build in R, and their parameters can be automatically determined. This is a critically important feature for rapid model development.

In the next chapter, we extend our use of the exponential type smoothing models into a powerful framework known as Component Form Exponential Smoothing.

Chapter 4

Using Component Form Exponential Smoothing

THE component form exponential smoothing approach is a flexible framework that considers variations in the combination of the trend and seasonal components. It includes all the traditional exponential smoothing models discussed in the previous chapter (e.g., Simple, Holt, Holt–Winters).

By the end of this chapter you will understand:

- The component form exponential taxonomy and why it is an easy to understand way to represent the components of time series data.

- Additional exponential smoothing methods of damped trends.

- Gain an intuitive understanding of information based model selection criteria.

- Use the component form to automate the forecasting of milk production using R.

The framework was originally developed to automatically forecast sales across thousands of pharmaceutical products. It has proven itself so useful that it was rapidly adopted into the business forecasting community.

Understanding Component Form Exponential Smoothing

An easy to understand way to represent the components of time series data is via the trend, seasonality, and the nature of the error. This is the approach take in the component form exponential smoothing framework. The component form representation is illustrated in the table below.

Error	Trend (T_t)	Seasonality (s_t)
Additive (A)	Additive (A)	Additive (A)
Multiplicative (M)	Multiplicative (M)	Multiplicative (M)
	Additive Damped (Ad)	No Seasonality (N)
	Multiplicative (Md)	
	No Trend (N)	

The above framework consists of 30 different exponential models (2 x 5 x 3 combinations). It breaks down the smoothing equations into their core components. These are the nature of the error, the trend component T_t and the seasonal component s_t. Each component form model has its own unique set of equations that consists of a forecasting equation and various smoothing equations similar to those we saw in the previous chapter.

> ***NOTE...*** ✍
>
> The error component is usually ignored in textbooks, because the distinction between additive and multiplicative errors makes no difference to the point forecasts.

Using the framework

The possibilities for each component are Trend $=\{A,Ad,M,Md,N\}$, seasonal $=\{A,M,N\}$, and Error $=\{A,M\}$. This allows us to represent any exponential model by its components. For example, Holt's method is denoted by (A,A,N) because it has additive errors, additive trend and no seasonality. The simple exponential model is characterized by (A,N,N). Similarly, (M,M,M) refers to a model with multiplicative errors, a multiplicative trend and multiplicative seasonality.

> ***NOTE...*** ✍
>
> Models with multiplicative errors are useful when the observations are positive.

Including Damped Trends

You may have noticed in the above table the additive and multiplicative damped trends. These components are useful when you need a short-term trend in your forecast, but require the long run forecast to be a constant. They are popular due to their remarkable forecasting performance. We discus these components in the next section.

Additive Damped Trends

The exponential models we have considered so far have a constant trend component. Forecasting a series into the future using a straight line is not necessarily appropriate as it can result in over-forecasting. To mitigate this a damping parameter ϕ can be used. In most circumstances we choose ϕ so that it lies between 0 and 1.

For example, the additive damped trend model is described by the smoothed series:

$$A_t = \alpha y_t + (1-\alpha)\left(A_{t-1} + \phi T_{t-1}\right) \tag{4.1}$$

with trend component:

$$T_t = \beta\left(A_t - A_{t-1}\right) + (1-\beta)\,\phi T_{t-1} \tag{4.2}$$

And forecast equation:

$$\hat{y}_{t+h} = A_t + \left(\phi + \phi^2 + \phi^3 + \phi^h\right) T_t \tag{4.3}$$

The dampening parameter, ϕ, controls how quickly the trend disappears over time. You might like to think of ϕ as a measure of the persistence of trends in the data. In fact, provided $0 < \phi < 1$, the long run forecasts (defined by letting $h \to \infty$) will be a horizontal straight line defined by:

$$\frac{A_t + \phi T_t}{1-\phi}$$

The practical implication of this is that short term forecasts will have a trend, and the longer-term forecasts will be constant. It makes the model extremely flexible in fitting the dynamics of a time series.

If $\phi = 0$, the method is identical to SEM, and for $\phi = 1$, it is identical to Holt's method. Furthermore, if $\phi > 1$, the forecast function has an exponential trend. This is useful if the empirical series has a strong trend.

NOTE... ✍

Damped trend predictions lie between those of SEM, and those of Holt's method.

Multiplicative damped trend model

As you might expect, we can also define a multiplicative damped trend model. The smoothed series equation is given by:

$$A_t = \alpha y_t + (1-\alpha) A_{t-1} T_{t-1}^{\phi} \tag{4.4}$$

with trend component:

$$T_t = \beta \left(\frac{A_t}{A_{t-1}} \right) + (1-\beta) T_{t-1}^{\phi} \tag{4.5}$$

And forecast equation:

$$\hat{y}_{t+h} = A_t T_t^{\left(\phi+\phi^2+\phi^3+\phi^h\right)} \tag{4.6}$$

The multiplicative model therefore involves modeling the trend by smoothing successive ratios of the local level and using a forecast equation that is the product of the level and growth rate.

Model Selection Criteria

Model selection is difficult. There is no consensus as to the right approach. There are numerous tools you can use. Two of the most popular are the Akaike Information Criterion and Bayesian Information Criterion.

Akaike Information Criterion

The Akaike Information Criterion (AIC) is a measure of the relative quality of a model based on Kullback-Leibler Information. We won't get into the details of Kullback-Leibler Information in this text. The key point is for each model (estimated on the same data) an AIC score is calculated. The model with the lowest AIC score is deemed to be the "highest" quality model. In other words, the "best" model is the one with the minimum AIC value.

AIC is therefore an estimate of relative quality. It offers a way to choose between competing models.

AIC is defined as:

$$AIC = -2 \ln L(\hat{\theta}) + 2p \tag{4.7}$$

where $\hat{\theta}$ is the maximum likelihood estimates of the model parameters, $\log L(\hat{\theta})$ is the corresponding log-likelihood, and $2p$ is a penalty on the log-likelihood as a function of the number of parameters p.

The method of maximum likelihood estimation is a very general approach to obtain estimates of the parameters of probability models. The idea is to choose the most likely values of the model parameters (i.e. θ), given the observed sample containing n observations $\{y_1, ...y_n\}$. Intuitively, the actual values these parameters take should depend in some way on the values observed in the sample data. This link is established via a probability model, which we denote by $f(x)$. The probability model is used to form the likelihood equation.

Statisticians figured out a while back that is easier to work in natural logarithms. Hence, the use of $\ln L(\hat{\theta})$ in equation 4.7.

Part of the reason for the popularity of AIC (among the academic community) is because it is an asymptotically efficient model selection criterion. As the sample gets larger (approaches infinity), the model with the minimum AIC score will be the highest quality (in fact it will possess the smallest Kullback-Leibler divergence). This also implies that for AIC to be valid, the sample should be large relative to the number of model parameters that need to be estimated.

The AICc metric is a popular refinement to AIC for small samples:

$$AICc = AIC + \frac{2p\,(p+1)}{n-p-1}$$

It tends to perform better than AIC if the ratio of the sample size to the number of parameters is small.

NOTE... ✍

Damped trend predictions lie between those of SEM, and those of Holt's method.

Bayesian Information Criterion

Bayesian Information Criterion (BIC) or Schwarz criterion is closely related to AIC, and they are often reported together. Similar to AIC, the BIC calculates a quality score for each model using the same data. The model with the lowest score is selected as the "best" candidate:

$$BIC = -2\ln L(\hat{\theta}) + p\ln(n)$$

Statisticians call BIC a "consistent" estimator. This means it will tend to select the 'true' model (if it exists) as the sample size increases (to infinity).

The BIC criteria reduce the risk of over-fitting relative to AIC because it has a penalty term ($p\ln(n)$) that grows with the number of model parameters. This is designed to filter out unnecessarily complicated models. In other words, BIC has preference for simpler models compared to AIC.

NOTE... ✍

The good news for us is that AIC and BIC are calculated automatically in R.

Advantages of Component Form Exponential Smoothing

The component form representation of the family of exponential smoothing models is very flexible because it extends to time series data that may exhibit:

- No trend, or alternatively exhibits an additive (linear) or multiplicative (non-linear) trend.
- Seasonal variation with additive or multiplicative patterns; or contain no seasonality at all.

It offers 30 separate models to choose from including all the traditional popular variants. As we see in the next section, the approach combined with automatic model selection, provides a powerful tool for efficient time series model construction.

Example - Forecasting Milk Production

Not too long ago, milkmaids in large numbers roamed the farms and rural countryside. In L. Frank Baum's book, *The Wonderful Wizard of Oz*, Dorothy trapped in a land of miniature china people encounters a milkmaid:

> *...and the first thing they came to was a china milkmaid milking a china cow. As they drew near, the cow suddenly gave a kick and kicked over the stool, the pail, and even the milkmaid herself, and all fell on the china ground with a great clatter.*
>
> *Dorothy was shocked to see that the cow had broken her leg off, and that the pail was lying in several small pieces, while the poor milkmaid had a nick in her left elbow.*
>
> *"There!" cried the milkmaid angrily. "See what you have done! My cow has broken her leg, and I must take her to the mender's shop and have it glued on again. What do you mean by coming here and frightening my cow?"*

> *"I'm very sorry," returned Dorothy. "Please forgive us."*
>
> *But the pretty milkmaid was much too vexed to make any answer. She picked up the leg sulkily and led her cow away, the poor animal limping on three legs. As she left them the milkmaid cast many reproachful glances over her shoulder at the clumsy strangers, holding her nicked elbow close to her side.*
>
> *Dorothy was quite grieved at this mishap.*

Today, the milkmaid is a relic of history, and milk production has been totally mechanized. In this section, we use component form exponential smoothing models to forecast average monthly milk production per cow in the United States.

Step 1 – Collect, Explore and Prepare the Data

The time series data we use in our analysis is contained in the TSA package. Load the data:

```
data("milk",package ="TSA")
```

The R object milk contains the observations. You can use the head function to see the first few observations:

```
head(milk)
```

```
       Jan  Feb  Mar  Apr  May  Jun
1994 1343 1236 1401 1396 1457 1388
```

So we see that for 1 January 1994 the value is 1343, for June 1994 the value is 1388.

Figure 4.1 plots the entire time series. Production rises gradually over the time period.

The monthplot function provides a simple way to visualize the differences by month in a time series. It collects together the data for each month into a separate time series:

```
monthplot(milk)
```

As can be seen from Figure 4.2, it allows you to visualize the underlying seasonal pattern and identify changes in seasonality by month. February is the lowest production month, and the series peaks in May.

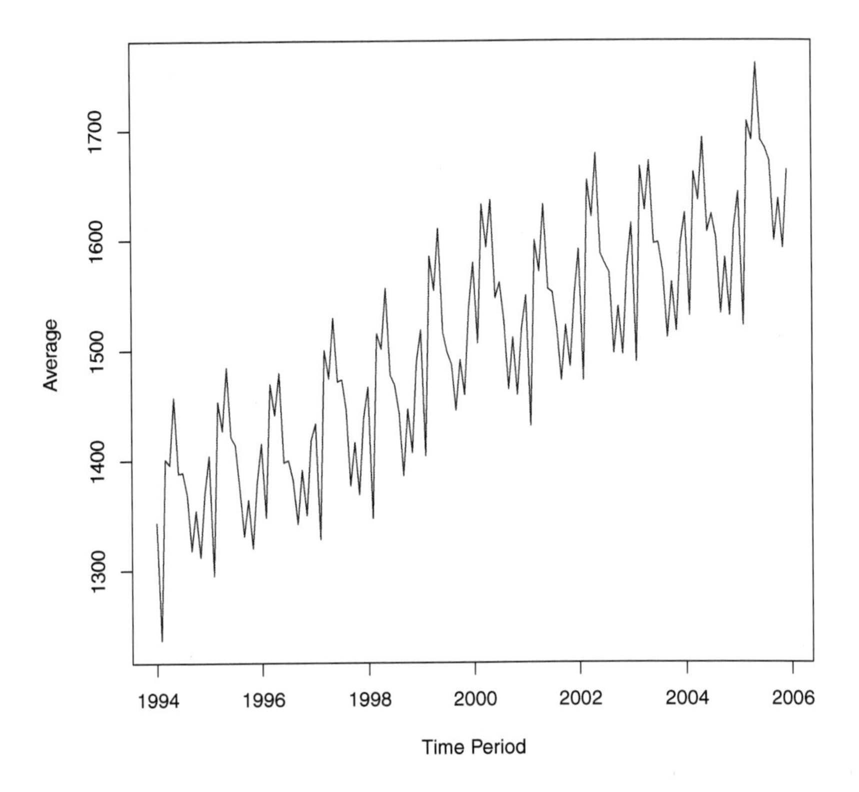

Figure 4.1: Average monthly milk production per cow in the US

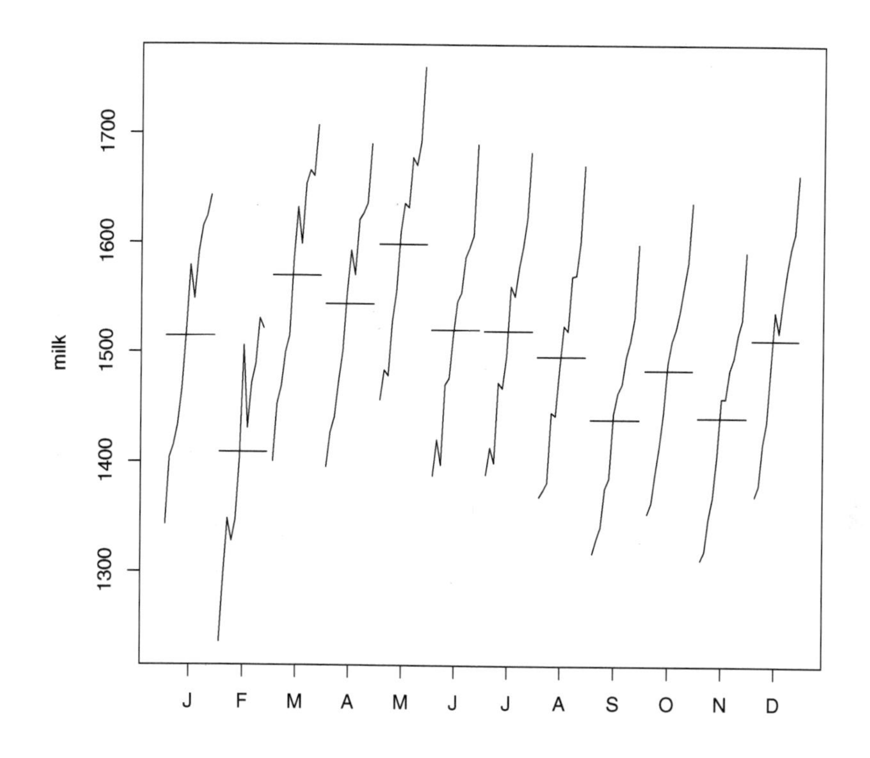

Figure 4.2: Monthly variation using `monthplot`

Step 2 - Build a Forecast Model

The first step is to choose a set of candidate models. These are typically derived from examining the characteristics of the time series visually, and/or from knowledge acquired from domain experts.

For this illustration we specify two models, the first is the simple exponential smoothing model. It has an additive error (Error = `A`), no trend (T_t= `N`), and no seasonality ($s_t = N$). It can therefore be represented as `ANN`.

The second candidate model is Holt's method denoted by

AAN. It has an additive error (Error =A), additive trend (T_t= A), with no seasonality ($s_t = N$). It can there be represented as AAN.

Neither of these models incorporate seasonality, an important component of the data. We look at how to capture this component automatically shortly, but first we walk through how to use the Exponential Component Framework with these two simple models.

Using the smooth package

We use the `smooth` package to build out models, specifying a model is straightforward:

```
require(smooth)
fit_es1<-es(milk,
model=c("ANN","AAN"),
h=12,holdout=TRUE,
intervals="parametric",
level=0.95)
```

Here are a few observations on the above R code.

- The fitted model is contained in the R object `fit_es1`.
- Candidate models are specified using the component form notation via the `model` argument.
- The last 12 months of data are held out for the test set, and we set a 95% parametric prediction interval.

Fitted values

The fitted and predicted values, for the optimal model, are automatically printed to the screen as shown in Figure 4.3. The optimal model is selected using the AIC. It is reported as AAN or Holt's method in Figure 4.3. As you might expect, it does a decent job capturing the trend, but is unable adequately capture the seasonality inherent in the data.

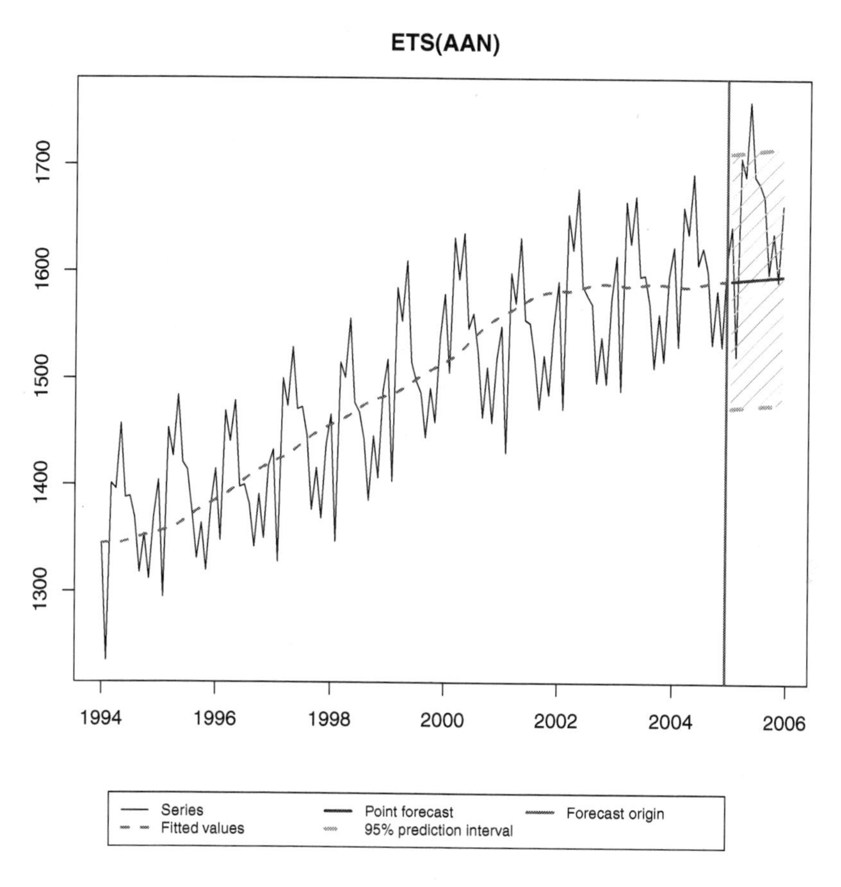

Figure 4.3: Optimal model from candidates in `fit_es1`

> ***NOTE...*** ✍
>
> You can set the `ic` parameter to select your model selection criteria. Set it to "`AIC`" for the Akaike information criterion (default), or "`AICc`" which corrects AIC for finite sample sizes. You can also set it to "`BIC`" to use the Bayesian information criterion.

You can also view the optimal model by appending the

$model argument:

```
fit_es1$model
```

```
[1] "ETS(AAN)"
```

Let's take a look at the smoothing parameters:

```
round(fit_es1$persistence ,4)
```

```
 alpha   beta
0.0055 0.0055
```

The estimated value of alpha and beta is 0.005. These values tell us that both the estimate of the current value of the level, and of the slope of the trend component, are based more on the smoothed values.

Step 3 - Evaluate Model Performance

We use MAPE as our assessment metric. You can view it for the fitted model as follows:

```
fit_es1$accuracy["MAPE"]
```

```
 MAPE
0.043
```

This value serves as a benchmark against which we can compare alternative models. We already suspect we should be able to improve on this number by incorporating seasonality into the model.

Step 4 - Improving Model Performance

Rather than manually selecting a model, you can let the algorithm choose for you. It will consider all 30 candidates in the component framework, and returns the model with the lowest AIC score. This is often the best choice, especially when you have numerous time series data for which models are required.

To use automatic model selection you simply pass the data as follows:

```
fit_es<-es(milk,h=12,
holdout=TRUE,
intervals="parametric",
level=0.95)
```

The only difference from the previous code is that we drop the `model` argument. This tells the algorithm to choose the best model among all possible candidates.

Figure 4.4 shows the fitted and predicted values. It informs us the optimal model is ANA, and as you can see it captures the trend, seasonality and level well.

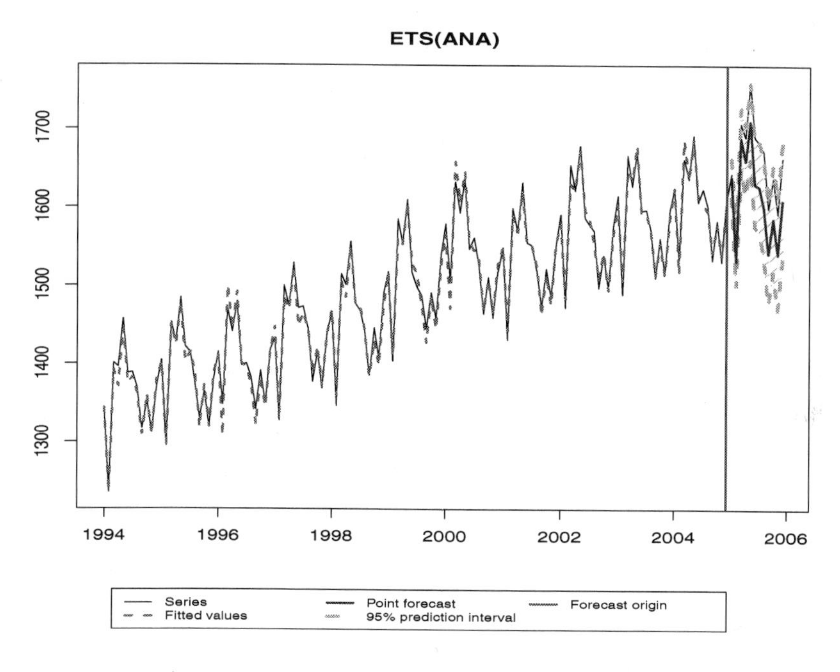

Figure 4.4: Automatic model selection using exponential component models

NOTE... ✍

Remember you can set the model selection criteria yourself by adding the `ic` parameter to the `es` function. It takes "`AIC`", "`AICc`" and "`BIC`".

Viewing MAPE

Now, take a look at the MAPE:

```
fit_es$accuracy["MAPE"]
```

```
 MAPE
0.026
```

At 2.6% it is considerably smaller than for the value reported for `fit1_es`. As we hinted at earlier, performance is enhanced in this example by including a seasonal component. The great thing here is that it was chosen for us automatically by the algorithm; this can speed up considerably the model identification, selection and deployment process.

My recommendation is to use this as your default choice when building exponential component type models; and only using the `model` argument when you have specific prior information that requires you to restrict the choices to a specific group/ type of models.

Limitations of Component Form Exponential Smoothing

Exponential smoothing models reduce some of the lag inherent to simple moving average models. However, like moving averages, are backward looking by nature.

They can also be sensitive as a large weight on recent observations tend to make the forecast whipsaw around. In this sense, the forecast can be more volatile than that of a SMA.

The automatic selection process is best suited for models that are required for short term forecasting. This is because several of the models in the exponential component framework can become unstable if used for long term forecasting.

Summary

Exponential smoothing models have delivered outstanding performance in a wide range of business areas. The component framework speeds up the entire forecasting process. Using this technique, hundreds of time series models can be tested and developed in a very short period. It is one of the key forecasting techniques in your time series prediction toolbox. You will return to this technique on numerous occasions.

In the next chapter, we look at another time series technique that has delivered outstanding performance for short term forecasting, the Theta method.

Chapter 5

Working with the Theta Method

SEVERAL years ago the Theta method for time series forecasting took the world by storm. It was shown to be very accurate in several prestigious forecasting competitions. The simplicity of the method and its stunning performance caught the attention of applied forecasters, who have used it with great success.

By the end of this chapter you will:

- Understand the core idea behind the Theta method.
- Clarify what theta lines are, and how they work.
- Walk through the steps in the standard Theta Method.
- Build Theta Method models using R.

The Theta methods ability to capture the dynamics of complex time series, and accurately project their values into the future was so astounding that I immediately added it to my time series forecasting tool kit.

Let's get started by taking a look at how this astonishing technique works, and then we'll get to play with it on some real data using R.

Understanding the Theta Method

The core idea behind the Theta method is the decomposition of a time series into two lines. One representing a long-term trend, and the other captures short-term dynamics. The lines are known as theta lines, and are used to predict the original time series. The model forecast is a weighted average of the predictions obtained by the theta lines.

Clarifying Theta Lines

The Theta method is based on the modification of the local curvatures of a time series through the Theta (θ) coefficient. A theta line is a weighted linear combination of the original time series data and a linear trend. For example, given the original time series y_t a theta line can be calculated as:

$$Z_t(\theta) = \theta y_t + (1-\theta)\left(\hat{\alpha} + \hat{\beta}t\right) \tag{5.1}$$

where $\left(\hat{\alpha} + \hat{\beta}t\right)$is the fitted time trend whose parameters $\hat{\alpha}$ (intercept) and $\hat{\beta}$ (slope) are estimated from the data.

To get a feel for these theta lines consider the case where $\theta = 0$. In this case using equation 5.1 we obtain:

$$Z_t(0)=\left(\hat{\alpha} + \hat{\beta}t\right)$$

This is the linear regression of the time trend. It essentially removes the curvature of the data to estimate the long-term trend component.

Other values of θ produce theta lines that are a weighted combination of the observed value y_t and the linear regression time trend. When $\theta > 1$, the local curvatures are increased, magnifying the short-term movements of the time series. When $\theta < 1$, the line approximates the long-term behavior of the series.

Theta lines have the same average value and slope as the original data. In other words, the modification through the coefficient θ retains the mean and slope of the original data but not their curvature. The local curvatures are diminished or enhanced, depending on the value of θ. The smaller (larger) the value of θ, the larger the degree of deflation (inflation). As we saw above, in the extreme case when $\theta = 0$, the time series is transformed into a linear regression trend line.

Each theta line is a separate time series. It can be extrapolated using an appropriate forecasting method. The separate forecasts are combined to calculate the point forecasts of the original time series.

Steps in the Standard Theta Method

The Theta method is a dynamic forecasting methodology. You can (in principle) choose how many theta lines to use. Then combine the individual forecasts with equal or unequal weights.

The Standard Theta Method (STM) uses two theta lines, with ad-hock values for θ, and equal weights for the combination of the final forecasts.

> ***NOTE...*** ✍
>
> STM is equivalent to simple exponential smoothing with drift.

The steps

STM consists of the following six steps:

Step-0: The time series is examined for seasonality using the autocorrelation function. A statistical test of the autocorrelation coefficients is used to determine whether seasonality is significant. For example, a time series contains seasonal behavior if:

$$|r_m| > q_{1-\frac{\alpha}{2}} \sqrt{\frac{1 + 2\sum_{i=1}^{m-1} r_i^2}{n}}$$

where r_k denotes the lag k autocorrelation function, n is the sample size, q is the quantile function of the standard normal distribution, $(1 - \alpha)$ the statistical confidence level, and m is the number of periods in the seasonal cycle, for example for monthly data $m = 12$, and for quarterly data $m = 4$.

Step-1: If seasonality is evident, perform a seasonal decomposition of the data assuming a multiplicative relationship for the seasonal component.

Step-2: The time series is decomposed into two lines, the linear regression time trend $Z_t(0)$, and the theta line $Z_t(2)$. Notice $Z_t(2)$ doubles the local curvatures of the series to approximate the short-term behavior.

Step-3: $Z_t(0)$ is extrapolated by linear regression, and $Z_t(2)$ is extrapolated using the simple exponential smoothing model.

Step-4: The forecasts produced from the extrapolation of the two lines are combined with equal weights (50%-50%).

Step-5: Finally, if the forecasts were deseasonalized in step 1, the seasonal component is added back.

Alternative Theta methods

Several alternative Theta models have been proposed:

- The optimized theta method (OTM) offers a slight variation on STM. In OTM $Z_t(2)$ is replaced by $Z_t(\hat{\theta})$ where the value of $\hat{\theta}$ is estimated directly from the data. This makes it more computational intensive than STM.

- The Dynamic Optimized Theta Model (DOTM) uses a state space model to select the best short-term theta line and dynamically revises the long-term theta. It is useful when you want to place greater emphasis on modeling the short-term behavior of a time series. When $\theta = 2$, the model is known as the Dynamic Standard Theta Model.

NOTE... ✍

In the OTM, $\hat{\theta}$ is obtained by minimizing the one step ahead prediction error.

Advantages of the Theta Method

The basic theta method is very simple to understand. It only uses two lines ($Z_t(0)$, and $Z_t(2)$) to generate forecasts. It does not require extensive training in statistics or detailed mathematical knowledge before it can be applied. It can be used on data with or without a trend, and whether seasonality is present.

Example - Forecasting Carbon Monoxide Concentrations

The yellow smog of industrial production is a common sight in cities around the globe. Pollution has long gone hand in hand with economic development. Richard Harding Davis's Victorian short story *In the Fog* conjurers up a striking image. In it, an American visiting London comments to a fellow traveler:

> "*You have never seen a London fog, have you?' he asked. 'Well, come here. This is one of the best, or, rather, one of the worst, of them.' I joined him*

> *at the window, but I could see nothing. Had I not known that the house looked out upon the street I would have believed that I was facing a dead wall. I raised the sash and stretched out my head, but still I could see nothing. Even the light of the street lamps opposite, and in the upper windows of the barracks, had been smothered in the yellow mist. The lights of the room in which I stood penetrated the fog only to the distance of a few inches from my eyes."*

Today environmental scientists measure Carbon monoxide along with other pollutants to assess the level of pollution. Carbon Monoxide (CO) is a poisonous, colorless, odorless, and tasteless gas.

In our cities, it is produced by fuel combustion in residential housing, businesses, industry and utilities. On our roads and highways, it is produced by cars, trucks, buses and vans. Environmentalists in the major industrial cities report it on a daily basis.

In this section, we use the Theta method to build models to forecast the monthly average concentration of CO in one of South Korea's largest cities - Gwangju.

Step 1 – Collect, Explore and Prepare the Data

Daily average CO concentrations for the city of Gwangju can be found in the `HEAT` package. Here is how to load the required data:

```
require(HEAT)
data("mort")
Gwangju <-  read6city(mort, 24)
```

The function `read6city` extracts a city-specific data set from the data frame `mort`. The code for Gwangju is 24, hence we pass that value as the second argument to the `read6city` function.

Take a look at Gwangju using the str function. It contains mortality, air pollution and, meteorological data:

```
str(Gwangju)
```

```
.frame':    2922 obs. of  24 variables:
 $ ccode     : int  24 24 24 24 24 24 24 24 24 24 ...
 $ cname     : Factor w/ 8 levels "","bs","dg","dj
    ",..: 5 5 5 5 5 5 5 5 5 5 ...
 $ yy        : int  2000 2000 2000 2000 2000 2000 2000
    2000 2000 2000 ...
 $ mm        : int  1 1 1 1 1 1 1 1 1 1 ...
 $ dd        : int  1 2 3 4 5 6 7 8 9 10 ...
 $ date      : Date, format: "2000-01-01" "2000-01-02"
    ...
 $ nonacc    : int  11 19 23 22 21 10 5 25 12 22 ...
 $ cardio    : int  3 7 6 8 4 1 3 12 4 5 ...
 $ respir    : int  0 1 4 3 2 1 0 3 2 2 ...
 $ influenza: int  NA NA NA NA NA NA NA NA NA NA ...
 $ meanpm10  : num  64.6 79.9 54.9 78.2 81.5 ...
 $ meanso2   : num  9.11 9.46 5.56 10.46 11.47 ...
 $ meanno2   : num  38.1 36.5 24 40.2 48.5 ...
 $ meanco    : num  10.1 13 5.3 11.1 10.5 ...
 $ maxco     : num  16 22 6.67 16.33 24.67 ...
 $ maxo3     : num  15 14 18.33 9.67 18 ...
 $ meantemp  : num  6.52 5.72 2.05 3.76 9 6.39 -0.82
    0.81 1.18 2.5 ...
 $ maxtemp   : num  12.7 7.8 5.2 11.3 14 9.8 1 4.8 3.1
    6.2 ...
 $ mintemp   : num  1.5 3 -1.3 -3.1 4.9 1.2 -2.5 -1.7
    -0.4 -0.9 ...
 $ meanhumi  : num  59.9 76.9 59.9 53.6 67 ...
 $ meanpress: num  1023 1018 1023 1023 1017 ...
 $ season    : int  4 4 4 4 4 4 4 4 4 4 ...
 $ dow       : int  7 1 2 3 4 5 6 7 1 2 ...
 $ sn        : int  1 2 3 4 5 6 7 8 9 10 ...
```

The variable we are interested in is the item meanco. It contains daily average CO concentrations from January 1st 2001 to December 31st 2007.

Working with dates

One of the most important parts of working with time series data involves creating derived time series. To do this effectively, it is critical to keep track of dates. To illustrate this idea, we create an R object to store dates used in our analysis. A neat way to do this is via the `seq` function:

```
date1 <- seq(as.Date("2000-01-01"),
as.Date("2007-12-31"),
by = "1 day")
```

The object `date1` now contains a sequence of dates starting at 1 January 2000 and ending on 31 December 2007.

We better check the dates are correct. First, use the `head` function to inspect the first few dates:

```
head(date1)

2000-01-01" "2000-01-02" "2000-01-03"
[4] "2000-01-04" "2000-01-05" "2000-01-06"
```

As we expected, the first date is the 1st January 2000, followed by 2nd January and so on.

Now, to inspect the later dates, you can do this using the `tail` function:

```
tail(date1)

[1] "2007-12-26" "2007-12-27" "2007-12-28"
[4] "2007-12-29" "2007-12-30" "2007-12-31"
```

Yep, it all looks good. The dates end December 31st 2007.

The next task is to combine the CO concentrations with the dates into a time series object. For this illustration, we use the `zoo` type. This can be achieved using the `zoo` function in the `xts` package:

```
require(xts)
co_con <- zoo(Gwangju$meanco, date1)
```

Now, take a look at the first eight observations in the R object co_con:

```
head(co_con,8)
```

```
2000-01-01 2000-01-02 2000-01-03 2000-01-04
     10.11      12.96       5.30      11.11
2000-01-05 2000-01-06 2000-01-07 2000-01-08
     10.48       6.41       5.87       7.28
```

The first reading reports a daily average CO concentration 10.11; and the eighth observation reports a value of 7.28.

The class function reports the nature of the co_con object:

```
class(co_con)
```

```
[1] "zoo"
```

And this is as expected.

> ***NOTE...*** ✍
>
> The **xts** and **zoo** objects are just two of the many types of objects that exist in R. The **xts** and **zoo** packages provide a set of powerful tools to make the task of managing and manipulating ordered observations fast and easy.

Dealing with missing values

Missing values are common with empirical data. People refuse or forget to answer a question, data is lost or not recorded properly. We all face the problem of missing data at some point in our work. Take another look at the output of the function str(Gwangju)listed above, did you notice the NA's in the influenza object? These are missing values.

We need to identify missing values (if any) in our object `code_con`. This is where the `sapply` function proves to be very useful. It is used to apply a function over a list or a vector. Simply pass `sapply` the relevant data, and a suitable function. Since we want the function to count the number of missing observations we pass it the data (`code_con`) and the `sum` function:

```
sapply(co_con,function(x)sum(is.na(x)))
```

```
[1] 35
```

It appears 35 observations are missing from the sample. This is not a large number given the sample size of 2,992 observations:

```
length(co_con)
```

```
[1] 2922
```

Nevertheless, we investigate a little further. The `imputeTS` package was designed to make it easy to visualize and deal with missing values. The `plotNA.distribution` function plots the time series observations with solid vertical bars at the location of each missing value:

```
require(imputeTS)
plotNA.distribution(co_con,
ylab="Daily average CO concentrations")
```

You should see Figure 5.1. The missing values appear to occur before 2004. There is a large solid block during 2000-2001; and a smaller block in 2003. What was the cause of this? Equipment failure, human error, who knows. The observations are missing and we have to decide what to do about them.

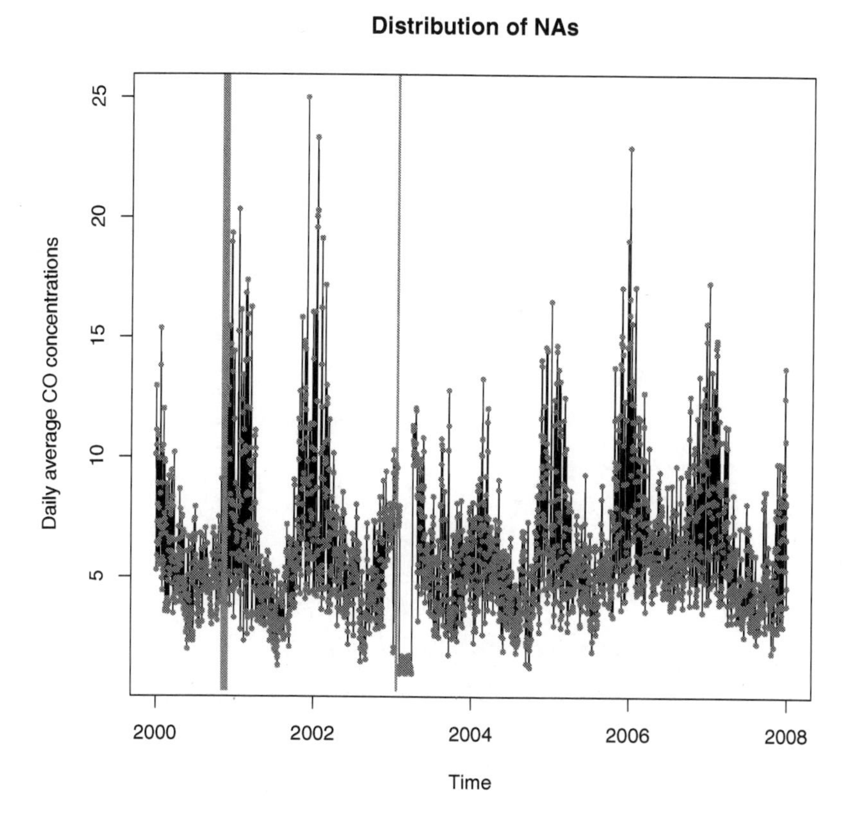

Figure 5.1: Time series plot of CO with missing values (vertical lines)

An alternative visualization, which you might find particularly useful, is the bar-plot. This is especially valuable for time series with a lot of observations. The `plotNA.distributionBar` is a neat little function that allows you to create a missing observation bar-plot quickly. The first argument takes the time series data, and the second argument the number of bins to be created. Here is an illustration with 100 bins:

```
plotNA.distributionBar(co_con,
breaks = 100)
```

As you can see, from Figure 5.2, the missing values are concentrated in two time periods. The first occurs around the 13% (of observations) mark, and the second around the 39% mark.

Figure 5.2: Bar plot of missing values

How should we deal with the missing values? The most common method and the easiest to apply is to use only those observations for which we have complete information. This is not appealing here given that we want to capture dynamics over time.

An alternative is to impute with a plausible value the missing observations. For example, you might replace the NA's with interpolated values. We will investigate this option using linear interpolation. The **na.interpolation** function in the **imputeTS** package allows us to do this. It takes two arguments, the first is the time series object; and the second is the type of interpolation:

```
clean_data <- na.interpolation(co_con,
"linear")
```

The object **clean_data** now contains the full data set with the missing observations replaced by the interpolated values.

Now take a look at the "cleaned" time series:

```
plotNA.imputations(co_con,
clean_data)
```

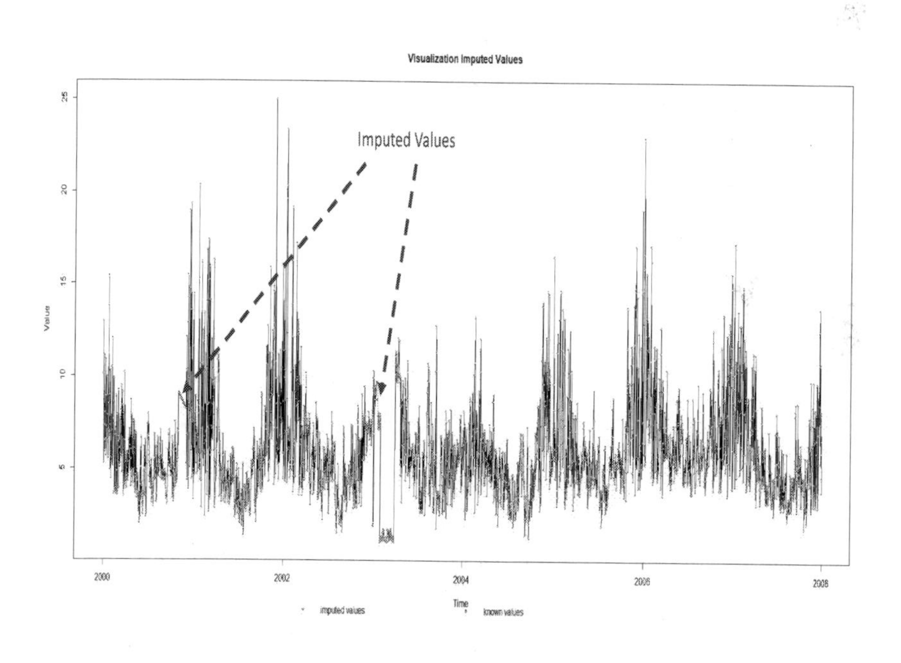

Figure 5.3: Imputed and original series using **plotNA.imputations**

The `plotNA.imputations` function visualizes the imputed values. Figure 5.3 shows an annotated version of the diagram.

Overall, the imputed values appear to fit the dynamics of the observed data, we therefore continue our analysis using these values in place of the NA's

> ***NOTE...*** ✍
>
> In the above illustration, we used linear interpolation. Other options include "`spline`" for spline interpolation; and "`stine`" for Stineman interpolation.

Transform daily data to monthly values

Our goal is to build a Theta forecasting model for the monthly average concentration of CO. Our data contains daily observations, we can convert these to monthly averages using the `apply.monthly` function from the `xts` package:

```
monthly.avg <- apply.monthly(clean_data,
   mean)
data_sample<-ts(monthly.avg,
frequency=12,
end=c(2007,12,31))
```

The monthly observations are stored in the R object `data_sample`. As shown in Figure 5.4, the average values show strong evidence of seasonality, and they appear to fluctuate around a constant mean level.

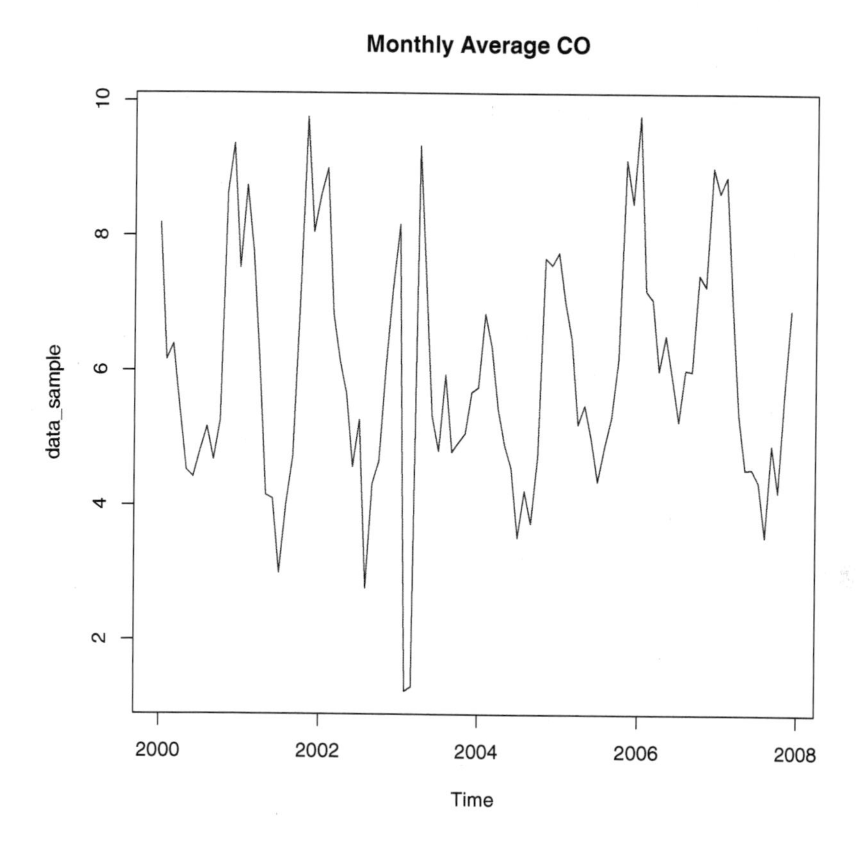

Figure 5.4: Average monthly values of CO for Gwangju

Step 2 - Build a Forecast Model

The forecTheta package contains routines for forecasting univariate time series using Theta models. The Dynamic Standard Theta Model can be fitted using the dstm function. For example, to forecast the next 12 months of CO average values with a 95% prediction interval you could use:

```
library(forecTheta)
fit_dstm<-dstm(data_sample,
level=c(95),
```

```
h=12)
```

The R object `fit_dstm` contains details of the fitted model.

The forecasts can be viewed by appending **`$mean`** to `fit_stm`:

```
round(fit_dstm$mean,2)

      Jan   Feb   Mar   Apr
2008 6.81 5.86 5.17 5.37
      May   Jun   Jul   Aug
2008 4.75 4.19 3.80 4.01
      Sep   Oct   Nov   Dec
2008 4.06 4.85 6.46 6.69
```

The first forecast, for January 2008, takes the value 6.81; and the forecast for December is 6.69.

Prediction intervals

You can also view the prediction intervals directly by appending `upper` and `lower` to the fitted model:

```
round(fit_dstm$upper,2)

        Jan    Feb    Mar
2008  10.36   8.87   7.77
        Apr    May    Jun
2008   8.59   7.38   6.96
        Jul    Aug    Sep
2008   6.07   6.88   6.53
        Oct    Nov    Dec
2008   8.19  11.11  11.80
```

The upper 95% prediction interval takes a value of 10.36 for January 2008, and 11.80 for December 2008. These values are somewhat higher than the point forecasts, and this is to be expected as they give us a sense (statistically speaking) of the expected upper limit of the predictions.

```
round(fit_dstm$lower ,2)

      Jan  Feb  Mar  Apr
2008 3.26 3.15 2.61 2.67
      May  Jun  Jul  Aug
2008 2.18 1.72 1.41 1.49
      Sep  Oct  Nov  Dec
2008 1.50 1.61 2.00 1.52
```

The lower 95% prediction interval for January is 3.26, and 1 .52 for December 2008.

Summing up, we see that the point forecast for February 2008 was 5.86, with an upper value of the 95% prediction interval of 8.87, and a lower value of 3.15.

Assessing normality

The prediction intervals are only valid (statistically) if the residuals are from the bell shaped normal distribution. We can visually assess the normality of the residuals using a density plot of the residuals, and a normal Q-Q plot:

```
par(mfrow=c(2,1))
plot(density(fit_dstm$residual),
main="Density plot")
qqnorm(fit_dstm$residual)
qqline(fit_dstm$residual)
```

Figure 5.5 shows the resultant plots. The multiple humps in the density plot (top figure) indicate the residual are non-normally distributed. In a Normal Q-Q plot the ordered residuals (y-axis) are plotted against the expected quantiles from a standard normal distribution function (x-axis). The plotted points should lie on an upward sloping (45 degree) straight line.

Figure 5.5 (bottom), shows that many of the residual points fall approximately along the straight upward sloping line. However, a rather large number in the right tail deviate considerably from the line.

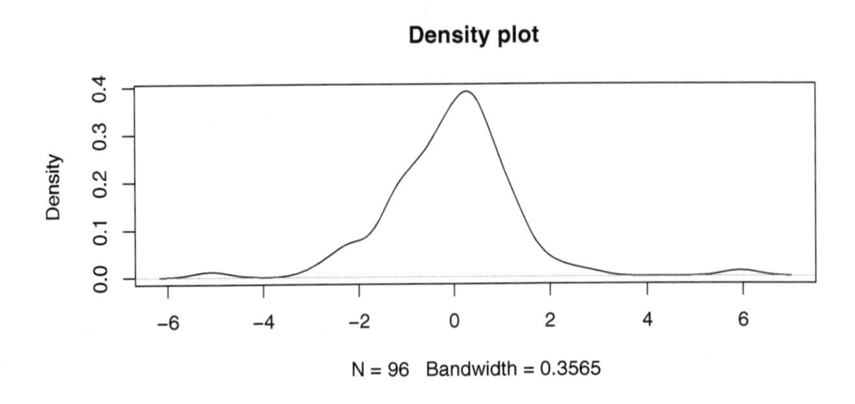

Figure 5.5: Density and Q-Q plot of residuals

Should we accept the assumption that the residuals are from the normal distribution? We can also use a formal statistical test to help us decide. The Shapiro-Wilks normality test is quick and easy to execute:

```
shapiro.test(fit_dstm$residuals)

        Shapiro-Wilk normality test

data:  fit_dstm$residuals
W = 0.92438, p-value = 3.466e-05
```

The null hypothesis of the Shapiro-Wilks normality test is that the residuals are from a normal distribution. The p-value of the test statistic, at 3.466e-05, is rather small (less than 0.001), and we reject the null hypothesis that the residuals are normally distributed. Therefore, we can't rely on the prediction intervals as being statistically valid.

> ***NOTE...*** ✍
>
> Remember, we reject the null hypothesis if the probability (`p-value`) is very small.

Step 3 - Evaluate Model Performance

We use the MSE as our assessment metric in this illustration. The `Metrics` package provides a simple way to calculate it. You simply pass the forecasts and observed values to the `mse` function:

```
n=length(data_sample)
end_train<-n-12
require(Metrics)
round(mse(data_sample[(end_train+1):n],
fit_dstm$mean),3)
```

```
[1] 1.535
```

The MSE is 1.535, we can use this value as a benchmark against which to compare alternative models.

> ***NOTE...*** ✍
>
> The objects `n` and `end_train` are used to align the observed values with the forecasts.

Step 4 - Improving Model Performance

Although the Standard Theta Method has been demonstrated to perform well with a wide range of time series, various adjustments have been proposed including the Dynamic Standard Theta Model. Whilst these alternative techniques can often perform very well, even outperforming the STM, it is always worth trying STM on your dataset. It can be called via the `stm` function:

```
fit_stm<-stm(data_sample,
level=c(95),
h=12)
```

The specification is identical to that of the `dstm` function. This makes the `forecTheta` package a delight to work with.

We assess the model using MSE, following similar steps to those outlined earlier:

```
round(mse(data_sample[(end_train+1):n],
fit_stm$mean),3)
```

```
[1] 1.529
```

This is a small improvement over the Dynamic Standard Theta Model. Alas, not large enough for us to get too excited, especially since it is measured off a few predicted values. Nevertheless, on the grounds of parsimony and performance STM wins out here.

Figure 5.6 shows the forecasts (dotted line) and actual values. We have not plotted the prediction intervals as the residuals are not normally distributed (you should check this for yourself). However, the forecasts, over this 12 month period, capture much of the dynamics of the series, especially in terms of the changing level over time.

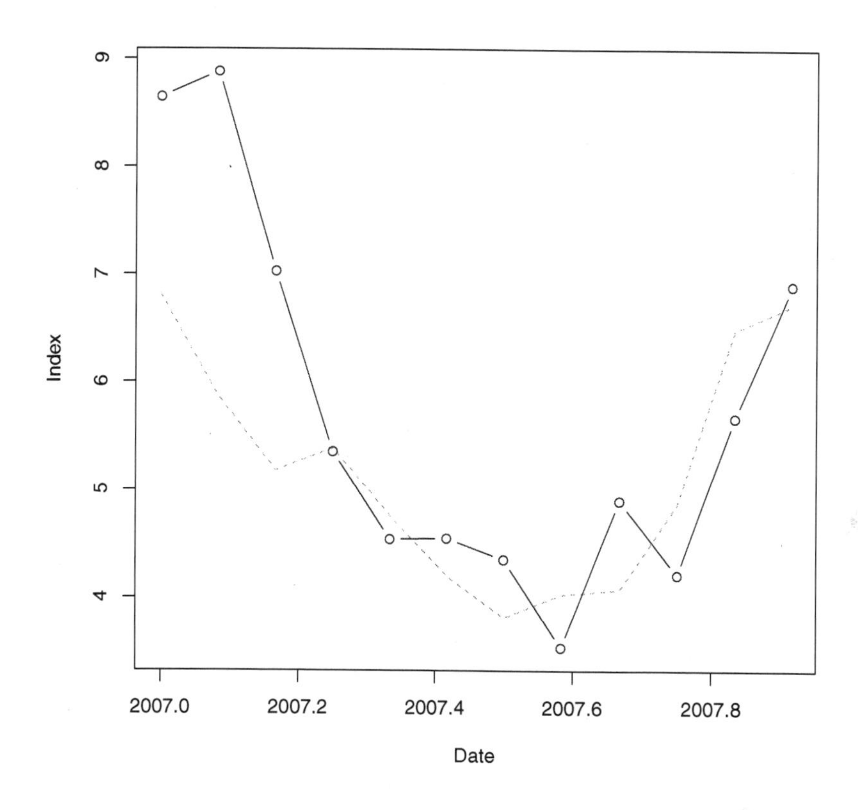

Figure 5.6: STM forecast and actual values

NOTE...

In the `forecTheta` package you can also call the optimized theta method via the `otm` function; and the Dynamic Optimized Theta Model via the function `dotm`. To try these functions simply pass your sample data followed by the forecast horizon (`h`).

Limitations of the Theta Method

An early criticism of the technique focused on the theoretical foundations. For example, the lack of statistical prediction intervals in the original method. In business, statistical prediction intervals are often less important than organizational tolerances for prediction error. These often drive rules of thumb for acceptable prediction ranges, and are often more useful than empirically derived statistical intervals.

Nevertheless, using the `forecTheta` package, you can, if you wish, easily generate prediction intervals for Theta models. Of course, as we did in this chapter you will want to formally assess the assumption of normality of the residuals is valid.

Summary

The Theta method is a robust technique that often performs very well on empirical data. It is based on the concept of modifying the local curvatures of the time series, obtained by a theta coefficient. In the standard approach, the time series is decomposed into at two theta lines representing the long term trend and short-term dynamics. The point forecast is a combination of the forecasts produced by the individual theta lines.

The method has provided outstanding performance in a number of formal forecasting competitions, and continues to be a popular technique in the business analyst's time series forecasting tool kit.

In the next chapter, we take a look at a more traditional statistical modeling tool, the auto regressive integrated moving average model.

Chapter 6

A Practical Introduction to ARIMA Modeling

ARIMA models are the most widely-used time-series methods found in business, health services, industry and economics. ARIMA stands for **A**uto-**R**egressive **I**ntegrated **M**oving **A**verage and is specified by three parameters: (p, d, q).

They were developed in the 1970s and are designed to describe, mathematically, the changes in a data series over time. The volume of practitioner and academic articles about ARIMA can be overwhelming. Indeed, you could spend your entire career building and refining this class of time series forecasting technique.

In this chapter, we break through the jargon, and by the end of it you'll understand:

- What the auto regressive component is and how it works.
- The core idea behind the moving average component.
- Seasonality and differencing, and their role in the ARIMA framework.

- Key steps to building effective ARIMA models.
- How to build automated and manual ARIMA models in R.

ARIMA Models are taught in economics classes, statistics courses, and even to MBA students who would like to use the technique to give them an edge in their professional career.

A large part of the attraction of the ARIMA approach is their exists a wide body of theoretical results that can be drawn upon to support the validity of a chosen model. ARIMA is an essential element in your time series toolkit. Let's get started.

Understanding the ARIMA Approach

ARIMA stands for **A**uto-**R**egressive **I**ntegrated **M**oving **A**verage and is specified by three parameters: ARIMA (p, d, q). We discuss each of the parameters in more details shortly.

For now, understand that the ARIMA approach to forecasting does not assume any particular pattern in the historical data of the series to be forecast. Instead, it uses an interactive approach to identify a potential model from a very general class of models.

In this section, we look at the:

- auto regressive component;
- moving average component;
- ARIMA model in full;
- role of the autocorrelation;
- steps to building effective ARIMA models,

NOTE... ✍

As with other modeling techniques, the chosen ARIMA model is assessed against historical data to see if it describes the series with sufficient accuracy.

Auto-regressive Component

The auto-regressive component is defined by:

$$y_t = \alpha + \phi_1 y_{t-1} + \phi_2 y_{t-2} + ... + \phi_p y_{t-p} + \varepsilon_t \tag{6.1}$$

where the residual term ε_t is random noise, assumed to be from the normal distribution with a mean of zero and constant variance equal to σ^2 (i.e. $\varepsilon \sim N(o, \sigma^2)$). The parameter α is a constant, and the $\phi_1, ..., \phi_p$ are the coefficients on past values of y_t.

Equation 6.1 is called an $AR(p)$ model. The parameter p is known as the order of the auto regressive process. For example, an autoregressinve model of order 2 ($AR(2)$) model would take the form:

$$y_t = \alpha + \phi_1 y_{t-1} + \phi_2 y_{t-2} + \varepsilon_t \tag{6.2}$$

This tells us that the present value of y_t is dependent on the previous two values (y_{t-1} and y_{t-2}) weighted by ϕ_1, and ϕ_2 plus a random error. For example, suppose that y_t with the stock market price of Facebook. Equation 6.2 tells us that today's price is dependent on the price yesterday, and also the price the day before that, plus a random error.

NOTE... ✍

The process of fitting an ARIMA model is also referred to as the Box-Jenkins methodology.

Moving Average Component

The moving average component uses past errors as explanatory variables. The observed series y_t is represented as a linear combination of previous error terms defined by:

$$y_t = \mu + \theta_1 \varepsilon_{t-1} + \theta_2 \varepsilon_{t-2} + ... + \theta_q \varepsilon_{t-q} + \varepsilon_t \quad (6.3)$$

where the order q determines the number of terms to include in the moving average model. For example, the $MA(1)$ model takes the form:

$$y_t = \mu + \theta_1 \varepsilon_{t-1} + \varepsilon_t$$

In this case only the past error impacts the current level y_t, this implies the process has a very short memory.

> ***NOTE...*** ✍
>
> The $MA(q)$ component is not technically a "moving average" because the θ_i parameters can be negative and don't necessarily sum to 1. The term "moving average" is used by convention.

Understanding Stationarity

We normally restrict AR, MA and ARIMA models to stationary data. A stationary time series is one whose statistical properties such as mean, variance, and autocorrelation, are all constant over time. The intuition behind this is that ARIMA models use previous lags of time series data to characterize future behavior. Modeling stable series with consistent properties involves less uncertainty and more accurate parameter estimates and forecasts. It is easier to forecast with data that is stable than data that is chaotic.

Differencing

If a time series is non-stationary you can make it stationary by differencing it. For example, if the time series y_t is observed to have a trend then you would use the difference:

$$\Delta y_t = y_t - y_{t-1} \tag{6.4}$$

You would then check to see whether the new series Δy_t is stationary. If it is then you would use this series as your data in the ARIMA model. If Δy_t is not stationary you would difference it again:

$$\Delta\Delta y_t = \Delta y_t - \Delta y_{t-1} \tag{6.5}$$

You would then check to see if $\Delta\Delta y_t$is stationary. If not, you would difference this series. The process continues until you have a stationary series, which you then use to build an ARIMA model.

> ***NOTE...*** ✍
>
> A stationary series has no trend in the data either upwards or downwards.

Differencing and ARIMA models

The "I" in ARIMA models stands for integrated. This refers to the number of times the time series was differenced in order to make it stationary. For example, in equation 6.4 the series was differenced once, hence $d = 1$ implying one difference; and in equation 6.5, $d = 2$.

Stationarity and the AR model

For stationary data, the AR model is fitted with a number of constraints on the parameters to ensure the stability of the forecasts. For example, for an $AR(1)$ the constraints take the form:

$$-1 < \phi_1 < 1$$

and for an $AR(2)$:

$$-1 < \phi_2 < 1$$

and:

$$\phi_1 + \phi_2 < 1$$

and:

$$\phi_2 - \phi_1 < 1$$

As the order of the auto-regressive process increases the constraints become more complicated. Fortunately, we don't have to worry about these because R takes care of them automatically.

Clarifying the ARIMA Specification

Recall that ARIMA stands for **A**uto-**R**egressive **I**ntegrated **M**oving **A**verage and is specified by three parameters- p,d and q.

We typically write an estimated model as ARIMA (p, d, q). Where p refers to the order of AR component, d to the number of times the data was differenced to make it stationary, and q refers to the order of the MA component.

ARIMA Models and Seasonality

The ARIMA framework can also handle seasonal data. Such models are written as ARIMA(p,d,q,) $(P,D,Q)_m$. The parameters (p,d,q) refer to the nonseasonal part of the model we have already discussed. The parameters $(P,D,Q)_m$ capture the seasonal components. The value m refers to the number of periods per season. For example, for quarterly data m=4, and for weekly effects in hourly data m=168.

A time series which has no trend has seasonality of period m when its average value varies in a cyclical pattern. For example, a daily time series with no trend has seasonality if the average values on different days of the week are different, but the average value for the same day of the week in different months is the same.

As an illustration, you might expect the number of sales of pizza pies on Friday's to be higher than the number of units sold on Monday's. You might end up estimating an ARIMA $(0,0,1)(0,0,1)_7$. This model has MA(1) terms at lags at 1 day for the actual series, and 7 days for the seasonality component.

NOTE... ✍

The capital P, D, and Q are the seasonal components of the AR, differencing, and MA components.

Simple illustration

Let's place the above in the context of ARIMA modeling. For simplicity, let D = 0, then we could write a pure seasonal ARIMA model as ARMA (P,Q):

$$y_t = \phi_1 y_{t-m} + \phi_2 y_{t-2m} + ... + \phi_P y_{t-Pm} + \\ \epsilon_t + \theta_1 \epsilon_{t-m} + \theta_2 \epsilon_{t-2m} + ... \theta_Q \epsilon_{t-Qm}$$

Seasonality implies the mean of the observations is not constant, but rather, fluctuates according to a cyclical pattern. For example, in a time series of cases of the flu, the average number of victims varies by season, but for the same month in different years we might expect a more or less constant value. This implies that seasonality is an additional cause of non-stationarity and should therefore be removed before applying the ARIMA methodology.

Seasonal differencing

If seasonality is observed in y_t we choose an integer such that D >0, to use seasonal differencing. This involves computing the difference between an observation and the corresponding observation in the previous season. For example, for first degree differencing we would use:

$$\Delta_m y_t = y_t - y_{t-m} \tag{6.6}$$

The series $\Delta_m y_t$, if stationary, can be used to estimate the parameters of the ARIMA model.

> ***NOTE...*** ✍
>
> In the full ARIMA(p,d,q,) $(P,D,Q)_m$ model, you might perform both regular differencing (Δ), and seasonal differencing (Δ_m) in order to create a stationary time series.

The Core Steps to Build Effective ARIMA Models

The ARIMA modeling process can get complicated. Fortunately, the fundamentals of effective design can be explained in three steps, all of which involve empirical analysis:

- The first stage involves the identification of an appropriate ARIMA model. Characteristics derived from the empirical data help you select an appropriate order.
- The second stage involves the estimation of model parameters. These are derived from the underlying data using an optimization algorithm such as the method of maximum likelihood. This stage occurs automatically within the R software environment. Your primary task is to review the estimated coefficients to ensure they match with prior expectations and reasonableness. This is often carried out using statistical tests, although subjective opinion may also be required.
- The third step investigates the model residuals. The goal is to assess whether they satisfy the necessary theoretical assumptions for the chosen ARIMA model to be valid. The primary focus in this stage is on the characteristics of the autocorrelation of the residuals.

Efficient Use of the Autocorrelation Function

In the ARIMA methodology there is considerable emphasis placed on the shape of the autocorrelation (and partial autocorrelation) function. This is because its characteristics determine the order of the AR and MA components of the model. For example, the correlation between observations h time periods apart for an AR(1) is:

$$p_h = \phi_1^h$$

This defines the autocorrelation function. Figure 6.1 visualizes it for positive and negative ϕ_1. Notice that:

- for a positive value of ϕ_1 (top), the ACF exponentially declines toward zero as the lag h increases.

- for negative ϕ_1 (bottom), the signs for the autocorrelation alternate between positive and negative. The ACF exponentially decreases to zero as the lag h increases.

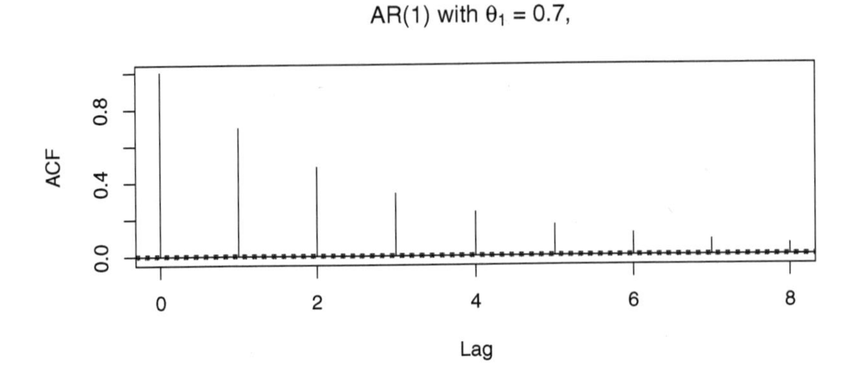

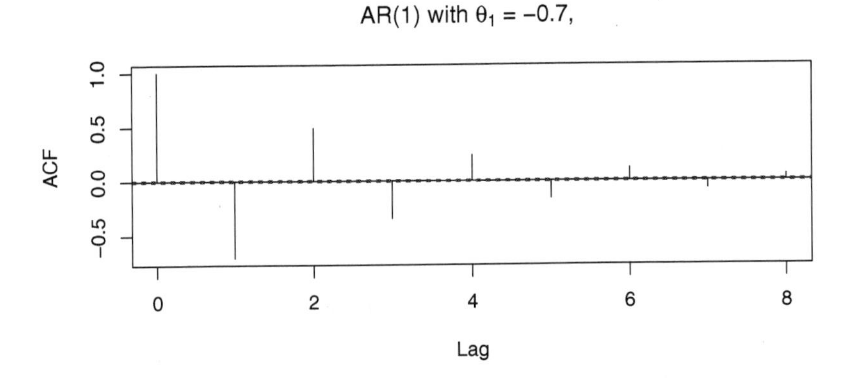

Figure 6.1: ACF for AR(1) model

ACF and MA models

For an MA (1) process the ACF is given by:

$$p_1 = \frac{\theta_1}{1+\theta_1^2}, \ and \ p_h = 0 \ for \ h \geq 2$$

For example, if $y_t = 2 + \varepsilon_t + 0.7\varepsilon_{t-1}$then:

$$p_1 = \frac{0.7}{1.49} \approx 0.47, \ and \ p_h = 0 \ for \ h \geq 2$$

As shown in Figure 6.2, the autocorrelation function spikes at the first lag with a value of 0.47.

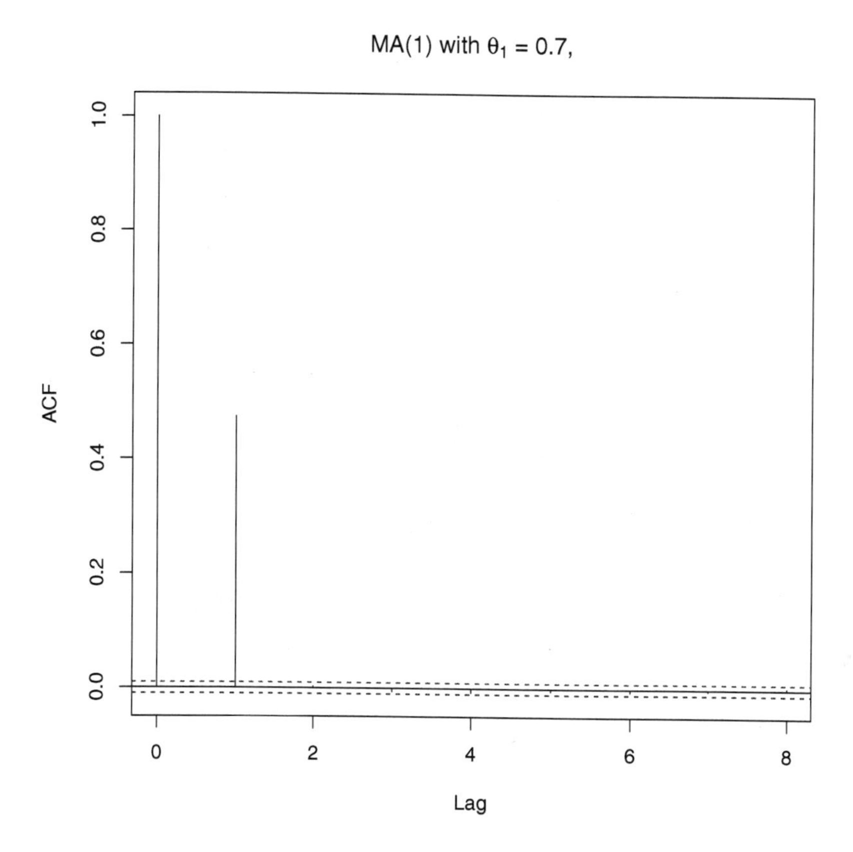

Figure 6.2: ACF for MA(1) model

As you might suspect an autocorrelation function can be defined for any AR or MA process. For example, for an MA(2) it is given by:

$$p_1 = \frac{\theta_1 + \theta_1\theta_2}{1 + \theta_1^2 + \theta_2^2},$$

$$p_2 = \frac{\theta_2}{1+\theta_1^2+\theta_2^2},$$

$$and\ p_k = 0\ for\ h \geq 3$$

Figure 6.3 plots the ACF for an MA(2) process where $\theta_1 = 0.7$ and $\theta_2 = 0.3$. The auto correlation for the first lag takes a value of approximately 0.58, and for the second lag a value of 0.19.

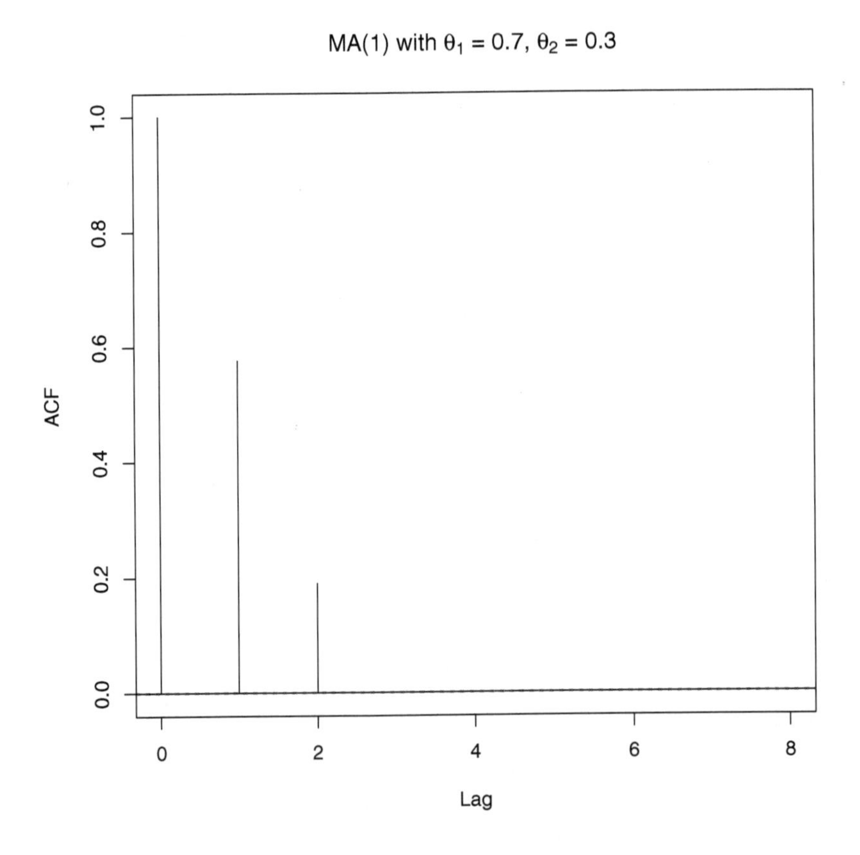

Figure 6.3: ACF for MA(2) model

Partial Autocorrelation function

The partial autocorrelation function (PACF) is useful for telling the maximum order of an AR process. It makes use of partial auto correlations. A partial autocorrelation is the amount of correlation between an observation x_t and a lag of itself (say x_{t-k}) that is not explained by correlations of the observations in-between. For example, if x_t is a time-series observation measured at time t, then the partial correlation between x_t and x_{t-3} is the amount of correlation between x_t and x_{t-3} that is not explained by their common correlations with x_{t-1} and x_{t-2}. The partial autocorrelation function (PACF) measures directly how an observation is correlated with an observation n time steps apart.

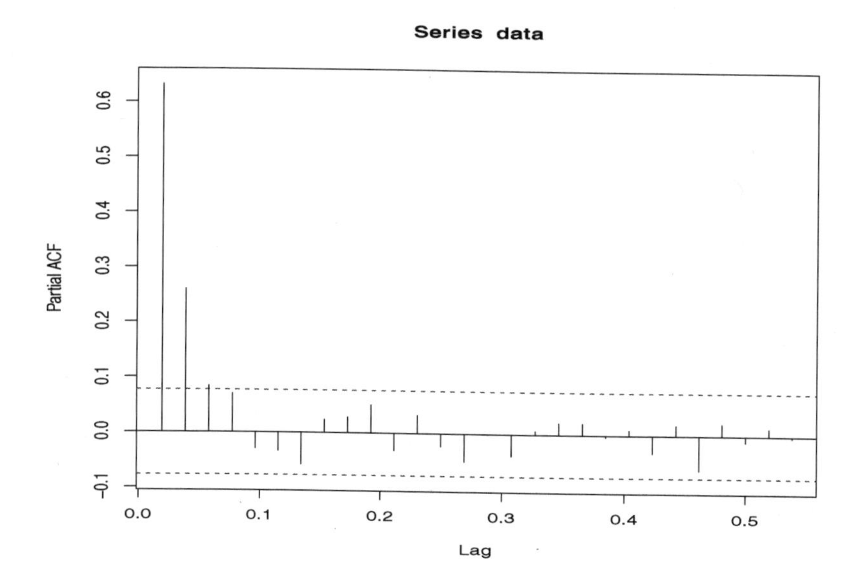

Figure 6.4: A sample PACF

Figure 6.4 shows a typical PACF. It plots the partial correlations at each lag. The interpretation of the PACF is somewhat subjective. We use it as a rough guide to the order of the model. For this example, the first four partial auto correlations are around or above 10%. We take this as an indication that using the past four observations might be a decent initial guess for the auto-regressive component.

Practice and theory

In practice, selecting the most appropriate model using the autocorrelation function (or partial autocorrelation function) is not as straightforward as the above theoretical illustrations. It involves subjective judgments about the characteristics of the autocorrelation function and partial autocorrelation function. In many cases, without having some experience with the data, it can be difficult to identify components such as seasonality trends or appropriate orders for the auto regressive and moving average elements.

A considerable volume of work has been produced in an attempt to generate simple rules of thumb. In practice, many of these rules are unusable for practical business purposes; or they require a level of expertise not generally available in large supply. This hampered the adoption of ARIMA type modeling in business and industry, and it was relegated to academia appearing in dusty scholarly journals until fairly recently. Fortunately, as we shall see shortly, we can use automatic procedures within R to help us out.

Advantages of ARIMA Models

Today, ARIMA type models are found in industry, business, health care research and government. They are founded on a mathematical model whose parameters are derived from empirical data. Like other forecasting methods discussed in this text ARIMA relies on the characteristics of the series being analyzed to project future period data. The estimated relationship can be extrapolated into the future. Seasonal characteristics can also be modeled if evident in the data.

Building ARIMA models can aid your understanding of the

properties that govern a specific time series. They can therefore be useful aids to the acquisition of domain expertise.

Example Forecasting Gun Sales

The citizens of the United States purchase more firearms than any other nation on earth. They collect them at antiques roadshow's, hoard them in giant gun vaults, shoot them in the movies and on the streets, and read about them in Western's and dystopian thrillers. *Murder in Gun-room* is a wonderful illustration of American's love affair with this deadly device. Here is the premise of H. Beam Piper's exciting book:

> *The Lane Fleming collection of early pistols and revolvers was one of the best in the country. When Fleming was found dead on the floor of his locked gunroom, a Confederate-made Colt-type percussion .36 revolver in his hand, the coroner's verdict was "death by accident." But Gladys Fleming had her doubts...*[this is] *a story that will keep your nerves on a hair trigger even if you don't know the difference between a cased pair of Paterson .34's and a Texas .40 with a ramming-lever.*

In this section, we build ARIMA models to forecast gun sales.

Step 1 – Collect, Explore and Prepare the Data

The data we use in our analysis is contained in the package and **gunsales**. Let's load the data, and take a look the column names via the **names** function:

```
require(gunsales)
gunsales <- analysis()
names(gunsales)
```

```
 [1] "year"                                   "month"
 [3] "guns_total"                             "guns_total_seas"
 [5] "guns_total_per_1000"                    "
guns_total_per_1000_scaled"
 [7] "handgun"                                "longgun"
 [9] "other"                                  "multiple"
[11] "longgun_share"                          "handgun_share"
[13] "new_jersey"                             "maryland"
[15] "georgia"                                "louisiana"
[17] "mississippi"                            "missouri"
[19] "dc_handguns_per_100k_national_sales"
```

The data frame contains information on the year and month of gun sales, the total gun sales and information on sales in various states, as well as information on the types of guns sold. We are interested in the total monthly sale of guns in the R object `guns_total`.

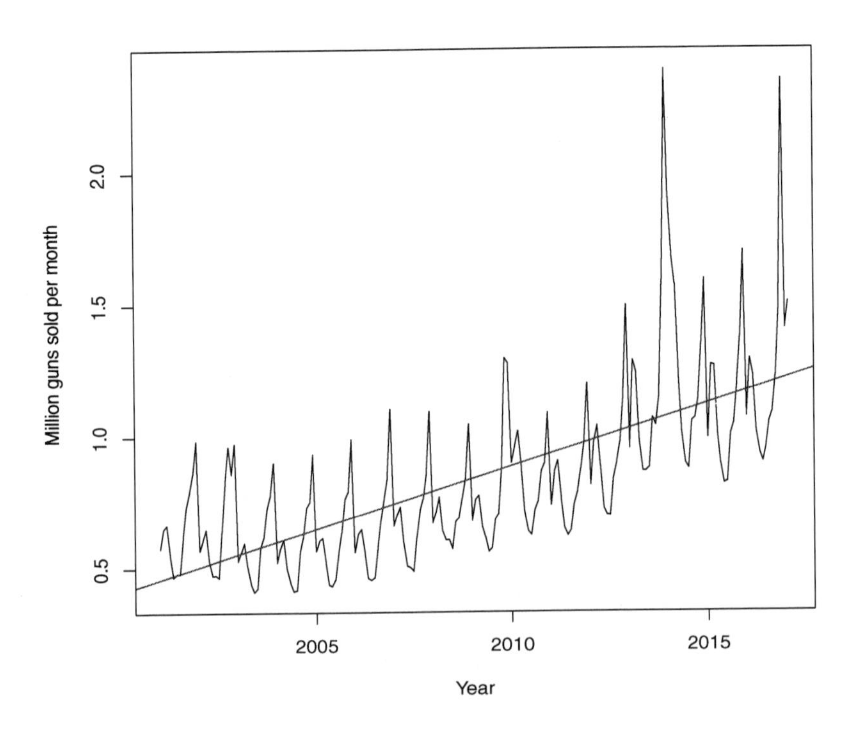

Figure 6.5: Total gun sales by month in the United States

Figure 6.5 shows the time series plot of the data, alongside a linear regression trend-line. A number of observations are immediately apparent from the visualization. First, there is a significant upward trend in sales over the time frame. Second, the data exhibits seasonality. And finally, the level of seasonality increases significantly towards the end of the time frame under consideration.

We get an even clearer picture using a multiplicative decomposition. This is illustrated in Figure 6.6. The change in trend becomes immediately more apparent, as does the seasonality. The random component appears to oscillate around zero.

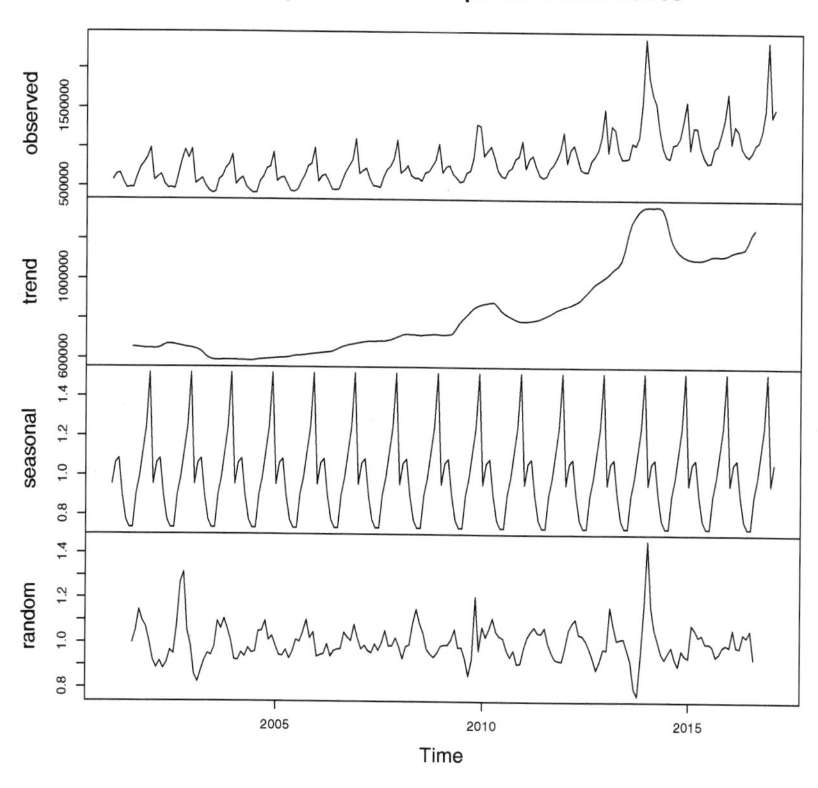

Figure 6.6: Decomposition of Gun sales data

An alternative representation of the seasonality is shown in Figure 6.7. It captures the characteristics by season using box plots. The average change by month is quite obvious, and there also appears to be some variation in the range of values as the year progresses. It appears June through August or possibly September has the lowest degree of variability in sales. Gun sales peak in December.

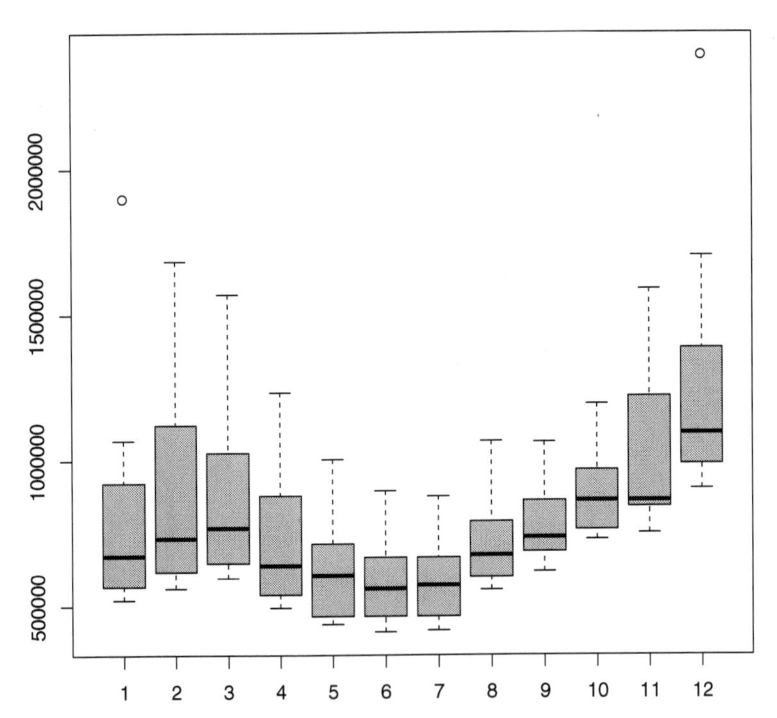

Figure 6.7: Box plot of gun sales by month

We build a forecasting model to predict the next twelve months of gun sales. The observations are stored as a time series object (ts) in `data_train`:

```
totalSales<-ts(gunsales[,"guns_total"],
start=c(2001,1),
frequency=12)
n=length(totalSales)
n_test=12
n_train=n-n_test
data_train<-ts(totalSales[1:n_train],
frequency=12, start=c(2001,1))
```

Step 2 - Build a Forecast Model

Earlier, we mentioned that the identification of the auto regressive and moving average components can be a challenge. Different analysts when faced with the same data may select different parameterizations. In addition, the interpretation of the autocorrelation function (or the partial autocorrelation function) can be fraught with difficulties and is inherently subjective. Fortunately, the selection of orders for the auto regressive and moving average component is automated in the `forecast` package. The function `auto.arima` will select both parameters via an information criterion (AIC, AICc or BIC). The specification is quite straightforward:

```
require(forecast)
fit<-auto.arima(log(data_train),
stationary=TRUE,
approximation=FALSE)
```

You can, if you want, simply pass the data to the `auto.arima` function. However, it has a number of useful features, two of which we illustrate in the above code. First, we set the `stationary` argument to `TRUE` to restrict the algorithm to search only for models which are stationary. Second, the `approximation` argument is set equal to `TRUE` to speed up calculation. Finally, you may have noticed that we take the logarithm of `data_train`, this sometimes helps with the normalization and linearization of the observations.

The object fit contains details of the fitted model. Simply type it into the console as follows:

```
fit

Series: log(data_train)
ARIMA(1,0,0)(0,0,2)[12] with non-zero mean

Coefficients:
         ar1     sma1     sma2     mean
      0.8131   0.7894   0.5950   13.5596
s.e.  0.0427   0.0799   0.0668    0.0963

sigma^2 estimated as 0.01226:  log
   likelihood=136.76
AIC=-263.52   AICc=-263.18   BIC=-247.5
```

Details of the fitted model are displayed on the screen. We see that the model has an AR (1) component, and a MA(2) seasonal component. The estimated value of ϕ_1 for the AR component is 0.8131, with the MA component $\theta_1 = 0.7894$, and θ_2 =0.5950. The output also reports the standard errors of the estimates (s.e.), alongside the log likelihood value and the AIC, AICc and BIC scores.

Remember, you can see the attributes of an object by using the attributes function. For example, to see those of fit you would type:

```
attributes(fit)

$names
 [1] "coef"        "sigma2"
 [3] "var.coef"    "mask"
 [5] "loglik"      "aic"
 [7] "arma"        "residuals"
 [9] "call"        "series"
[11] "code"        "n.cond"
[13] "nobs"        "model"
```

```
[15] "bic"          "aicc"
[17] "x"            "fitted"

$class
[1] "ARIMA" "Arima"
```

Let's take a look at the first few fitted values using the **head** function:

```
head(round(fit$fitted,2))

         Jan   Feb   Mar   Apr   May   Jun
2001 13.38 13.33 13.41 13.37 13.21 13.12
```

So, for January we observe a value of 13.38, and for June a value of 13.12. Now, remember these are the log values, to see the observations in the original scale use the **exp** function as follows:

```
head(round(exp(fit$fitted),0))

         Jan    Feb    Mar    Apr    May
2001 646255 616745 665386 638980 546937
         Jun
2001 499441
```

Figure 6.8 shows a plot of the fitted and actual values. Notice how well the model captures the dynamics of the underlying observations. This is quite common in ARIMA type models because of the flexibility inherent in their mathematical form. It also makes them very sensitive to over fitting, and we'll need to pay special attention to this risk. Over fitting occurs when your model perfectly fits the training data, but fails miserably to predict future (unseen) observations.

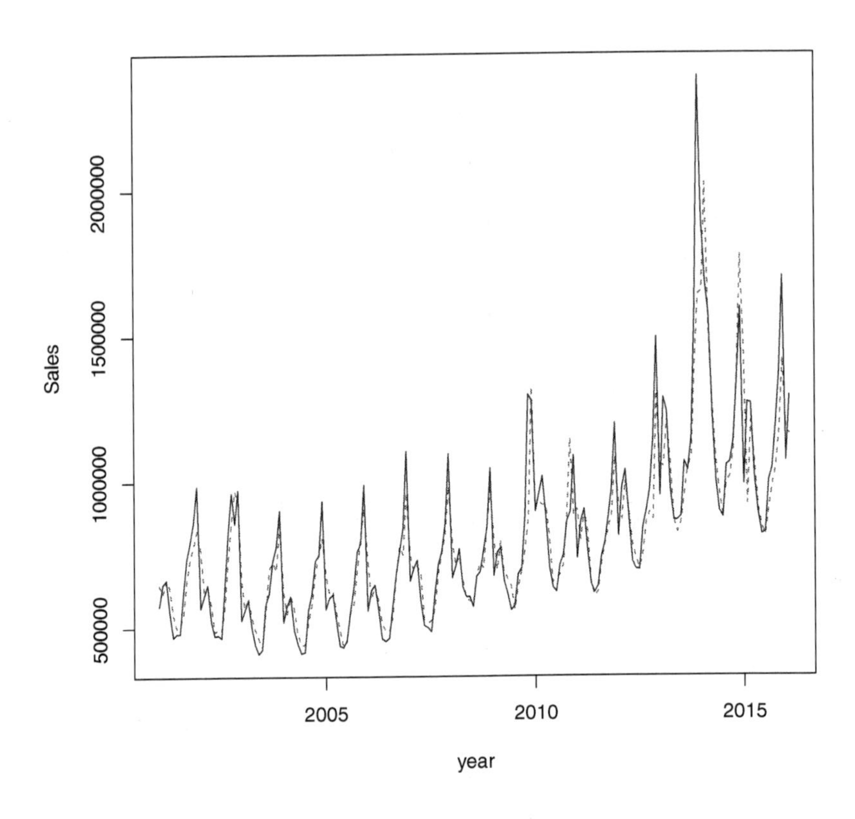

Figure 6.8: Fitted and actual values for gun sales data

Step 3 - Evaluate Model Performance

ARIMA models are a traditional statistical technique in the sense that they depend on a number of statistical assumptions. Two of the most important are that the residuals are random, and they are normally distributed. We can assess the validity of these assumptions using the `checkresiduals` function:

```
checkresiduals(fit)
```

Figure 6.9 shows the output. The top chart presents the residuals, they appear to be random and oscillate around zero.

A density plot is also reported alongside a solid curve representing the normal distribution. Visual inspection indicates the residuals are approximately normally distributed. Finally, the autocorrelation function indicates significant autocorrelation at twelve, twenty-four, and thirty-six months. This suggest there may be some residual seasonality not picked up by the model.

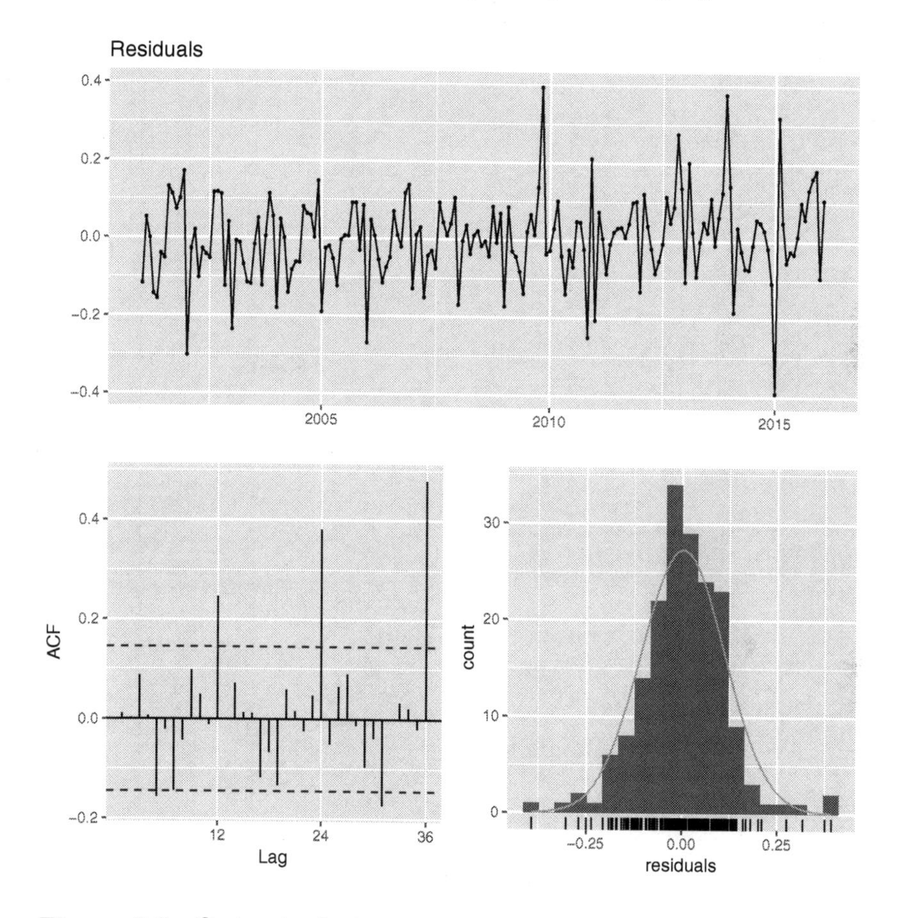

Figure 6.9: Output of `checkresiduals` function for gun sales model `fit`

Now, let's take a look at how the model performs on the test sample. We can use the `forecast` function to generate the predictions:

```
pred<-forecast(fit,
h=12,
level=0.95)
```

We set the parameter h =12 to generate twelve monthly predictions, and the prediction interval set to the 95% level via the `level` argument.

How well the model fits the data can be assessed using the root mean squared log error function (**rmsle**) in the `Metrics` package:

```
data_test<-ts(totalSales[(n_train+1):n],
frequency=12, end = c(2017, 2))
require(Metrics)
fmean<-exp(pred$mean)
round(rmsle(fmean,data_test),4)
```

```
[1] 0.319
```

So we see that the root mean squared log error takes a value of 0.319. We'll use this value as a benchmark against which to compare alternatives.

Figure 6.10 plots the forecast values (light gray line), the 95% level prediction intervals (dotted gray line), and the actual observed observations. The forecasts capture much of the observed dynamics of the data, however the last three actual values are outside the prediction interval.

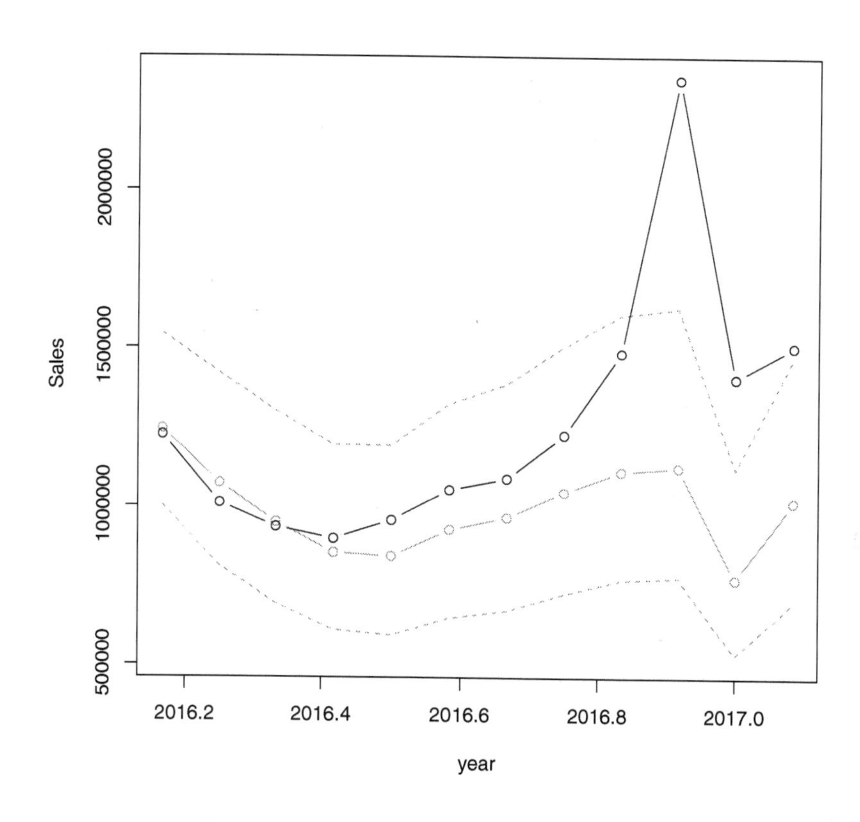

Figure 6.10: Test sample forecasts and actual observations

Step 4 - Improving Model Performance

ARIMA models were originally designed for the analysis of stationary data. As you can see from Figure 6.8 the time series has a pronounced upward trend. We saw earlier that it is possible to make the time series stationary by differencing it. The R function **ndiffs** is useful because it will tell you how many times you need to difference a time series to make it stationary:

```
ndiffs(data_train)
```

```
[1] 1
```

In this case a single difference is sufficient.

> ***NOTE...*** ✍
>
> The `ndiffs` function uses a statistical hypothesis to determine the number of differences required to make a time series stationary. The procedure it uses is known as a unit root test.

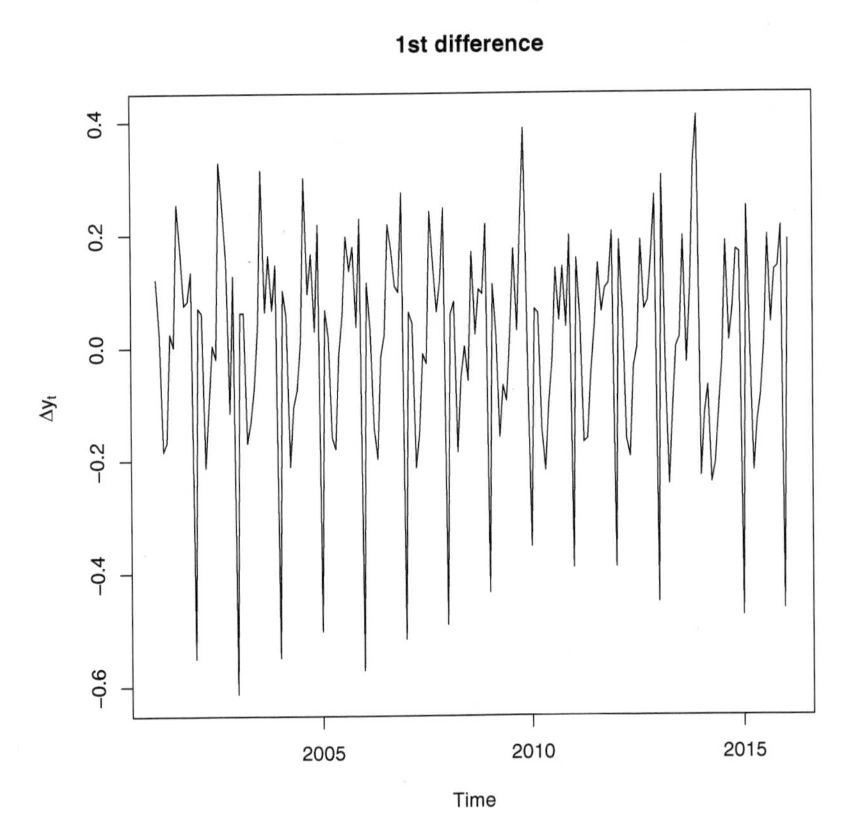

Figure 6.11: Differenced series

Figure 6.11 shows the differenced series. The trend has been removed, however seasonality is still present in the series. Of course, we can also difference the data to remove it (you should do this as an exercise).

Now, let's use the knowledge we have gained about the dynamics of the data to manually build an ARIMA (1,1,1) $(1,0,1)_{12}$:

```
fit1<-Arima(log(data_train),
order=c(1,1,1),
seasonal=c(1,0,1),
include.constant =TRUE)
```

Again, we pass the log of the data followed by the order and seasonal components. By setting `include.constant = TRUE`, we force the model to include a constant. Depending on the speed of your machine there may be a slight delay as the model is optimized. The optimized model is stored in `fit1`:

```
fit1
```

```
Series: log(data_train)
ARIMA(1,1,1)(1,0,1)[12] with drift

Coefficients:
          ar1      ma1    sar1     sma1   drift
       0.8681  -1.0000  0.9936  -0.6889  0.0038
s.e.   0.0387   0.0032  0.0038   0.0621  0.0012

sigma^2 estimated as 0.004965:  log likelihood=210.72
AIC=-409.44    AICc=-408.96    BIC=-390.25
```

The model specification is reported alongside the parameter estimates. The auto regressive component takes a value of 0.8681, and the seasonal auto regressive component takes a value of 0.9936 (almost 1). The moving average component takes a value of exactly -1, with a seasonal coefficient close to +1 (0.9936).

Log likelihood and residuals

The log likelihood at 210 is higher than that of our earlier model (`fit`), this is sometimes taken as an indication of a poorer fitting model. However, I've found that this is not necessarily the case in applied practice. It is often more useful to compare models directly on their performance with the test sample, and we'll get to that in a moment, but first let's use the **`checkresiduals`** function to assess the properties of the residuals:

```
checkresiduals(fit1)
```

Figure 6.12 shows the resultant plot. From the top diagram, which shows a time series of the residuals, we can see that they are distributed randomly around zero. The density plot also indicates normality of the residuals, at least visually.

In terms of the auto correlation's, two things are immediately evident relative to Figure 6.9. First, the seasonal spikes in the autocorrelation function are no longer evident, indicating that particular seasonal component has been effectively removed. Second, the auto correlation's (except for one value) lie within a 95% confidence interval (dotted line). This indicates that the model is capturing much of the characteristics of the underlying data.

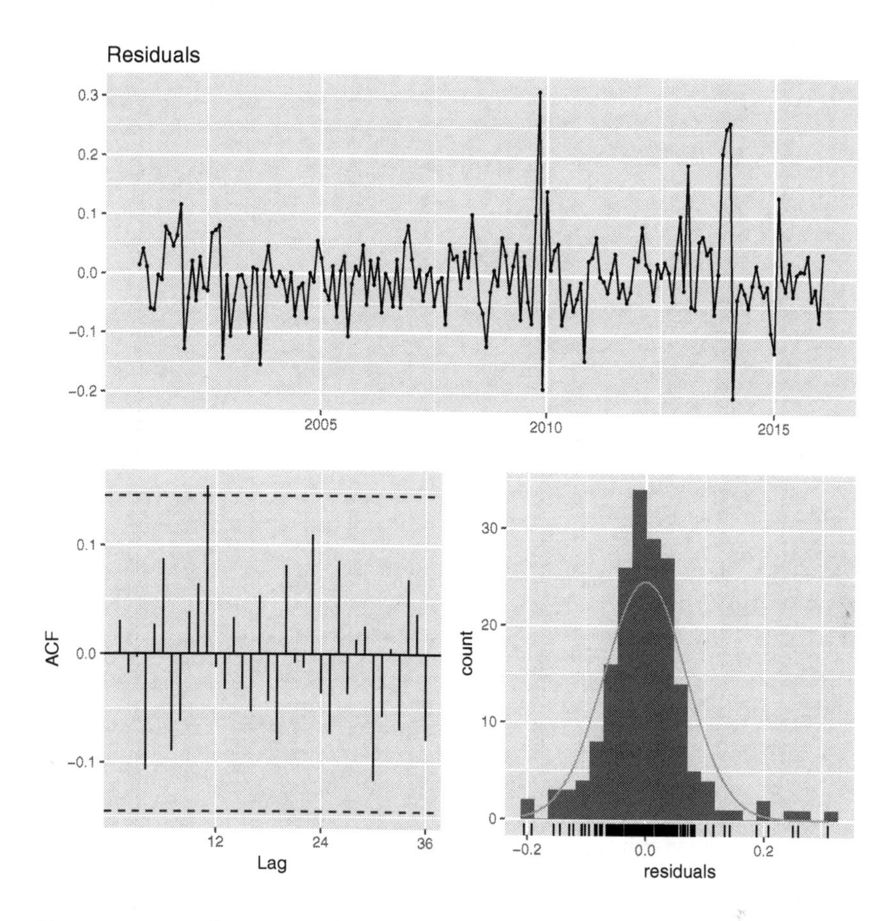

Figure 6.12: Output of `checkresiduals` function for gun sales model `fit1`

Overall, we might conclude that the characteristics of Figure 6.12 represent an improvement over those displayed in Figure 6.9. Nevertheless, we need to assess the performance of this model on the test data. As we did previously, we can achieve this using the `forecast` function:

```
pred1<-forecast(fit1,
h=12,
level=0.95)
```

The R object `pred1` contains the predicted values alongside the prediction intervals. The key question is whether `fit1` is better suited than `fit` given our forecasting problem and the underlying characteristics of the data.

> ***NOTE...*** ✍
>
> Calculate the `rmse` for this model and compare it the earlier model, what do you observe?

Visual assessment

One way to assess this is visually, and Figure 6.13 shows the predictions alongside the actual observations in the 95% prediction interval. We see that the actual observations are well contained within the prediction intervals, and in this sense `fit1` has the edge over the automatically selected model fit.

It is good to remember that the automated procedures are very useful for generating models, and in very many cases you may want to take the models that are generated from the `auto.arima` function and fine-tuned them for your specific needs.

Selecting a specific ARIMA model from scratch is somewhat subjective and the reliability of the chosen model can depend on the skill and experience of the forecaster. However, this problem can be reduced considerably by the use of automatic model selection algorithms like those discussed in this chapter.

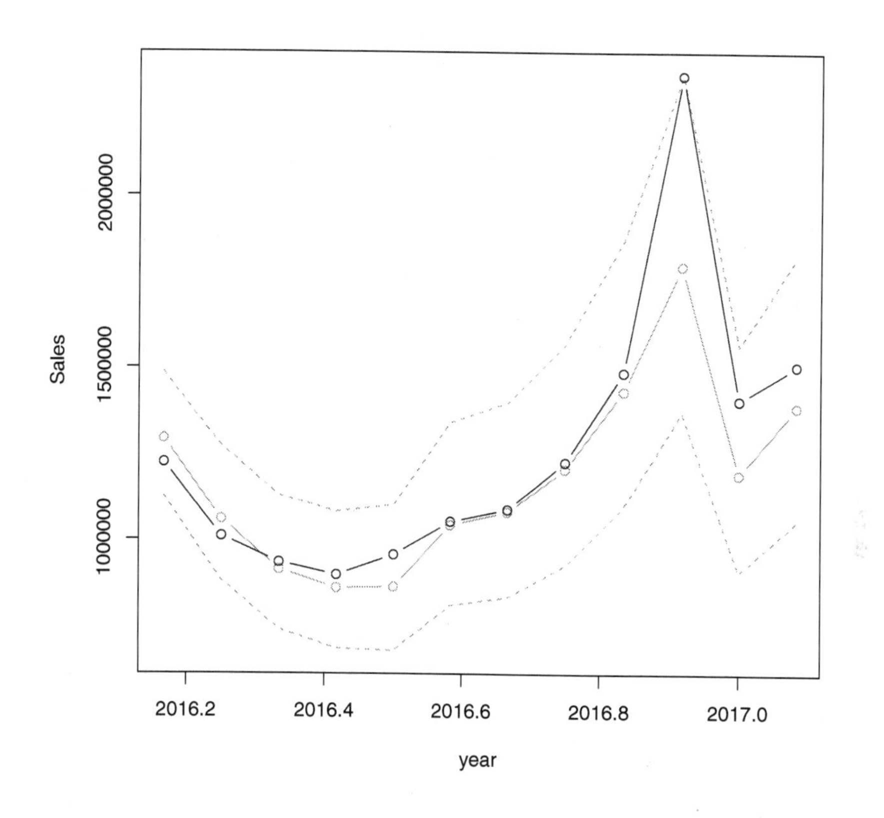

Figure 6.13: Test sample forecasts and actual observations for `fit1`

> ***NOTE...***
>
> With non-stationary data, the ARIMA model is useful for short-term forecasts only. If the actual data fall outside the prediction level of the forecast value, the model should be updated immediately.

Limitations of ARIMA

ARIMA model parameters are estimated directly from the data. They work well on long, stable, time series. In this situation, the prediction accuracy of the ARIMA model is often very high. The flexibility of the mathematical form of ARIMA type models make them susceptible to over fitting. This risk can be diminished somewhat by the use of a training and separate test sample. In many cases a separate validation sample is also used as a final test of the effectiveness of the selected model; and this is good practice, one you should adopt in your own analysis. However, as with other types of model, the forecast error will gradually increase as the number of prediction periods rises.

Summary

ARIMA Models have been used successfully for several decades to predict a wide variety of time series data. Although they require a higher degree of statistical knowledge than many of the other techniques discussed in this text, their accuracy and performance make them worth the investment. They can be trained, built and selected automatically using R. For this reason, we add them to our box of automated time series forecasting techniques.

Of course, a detailed understanding of statistics and mathematics is not required to develop highly accurate prediction models. As you'll see in the next chapter, intuitively easy to understand and apply techniques often perform extremely well.

Chapter 7

Mastering BATS and TBATS Forecasting Techniques

TRADITIONAL statistical techniques for time series forecasting are not well-suited for dealing with seasonality at multiple levels. In other words, where seasonality occurs simultaneously within a hour, day and week etc. BATS, and an extension called TBATS were developed specifically to address such problems. They decompose a time series to capture the trend and seasonal factors which can include long-run cycles, monthly seasonality, day of week effects and so on.

In this chapter, you will:

- Gain knowledge of the components of a BATS model
- Clarify the distinction between BATS and TBATS.
- Learn about the Mean Absolute Scaled Error, and discover a simple way to interpret it.
- Build BATS and TBATS models efficiently in R.

BATS and TBATS models make it possible to forecast data with more than one type of seasonality. This can enhance prediction accuracy, and increase the likelihood of "buy-in" from potential stakeholders. Let's dig into this powerful set of forecasting tools.

Understanding BATS and TBATS

We have already seen that the classical decomposition model for a time series y_t can take the form:

$$y_t = \begin{cases} trend + seasonality + error & [additive\ model] \\ trend \times seasonality \times error & [multiplicative\ model] \end{cases}$$

In practice, a time series may have many levels of seasonality. We can capture these with the factor s_i, where $i = 1, ..., T$ refers to a total of T-levels of seasonality in the time series. For example, if $T = 4$, s_1 might capture a pattern repeating every year, s_2 a pattern repeating every month, s_3 weekly seasonality, and finally s_4 might capture daily seasonality. The BATS and TBATS models are capable of handling such multiple levels of seasonality.

BATS Model

The BATS model is given by a series of equations. These can seem quite complex on first pass. In this section, we break each equation down, step by step, to clarify their role.

Box- Cox Transformation

The first step transforms the original data y_t through:

$$y_t^{(\omega)} = \begin{cases} \frac{y_t^{(\omega)}}{\omega}, & \omega \neq 0 \\ \ln(y_t) & , \omega = 0 \end{cases} \tag{7.1}$$

Equation 7.1 defines a family of transformations which depend on the parameter ω. They are known as a Box-Cox transformation.

The Box - Cox transformation was designed to modify the distributional shape of data so it would be "more" normally distributed. In other words, it is a way to transform a non-normal distributed variable into a normally distributed variable.

Statisticians use the technique because normally distributed data allows numerous statistical tests and procedures. In statistical practice, ω values of -2, -1.5, -1, -0.5, 0, 0.5, 1, 1.5, and 2 are typically investigated to determine which, if any, is most suitable to normalize observations.

The optimal value of ω can also be estimated directly from the data using an optimization technique such as the method of maximum likelihood.

The key thing to take away is that the variable $y_t^{(\omega)}$ is simply the Box Cox transformation of y_t.

NOTE... ✍

The Box- Cox transformation was developed in the early 1960s when Statistician Sir David Cox visited his British friend Professor George Box in Wisconsin. They decided to collaborate on a research paper.

Relationship to level, trend and seasonality

The BATS model employs a slight modification of the Holt-Winters method where the Box-Cox transformed observations are related to the level, trend and seasonality by the equation:

$$y_t^{(\omega)} = l_{t-1} + \phi b_{t-1} + \sum_{i=1}^{T} s_{t-m_i}^{(i)} + d_i \qquad (7.2)$$

where:

- l_{t-1} is the level of the series in time period t-1 ;
- b_{t-1} the trend with damping constant ϕ;

- The constants $m_1, m_2, ..., m_T$ are the seasonal periods, with $s^{(i)}_{t-m_i}$representing the ith seasonal pattern for period $t - m_i$;

- and d_i is an auto regressive moving average process [ARIMA (p,q)].

Level Equation

The equation defining level is given by:

$$l_t = l_{t-1} + \phi b_{t-1} + \alpha d_t \tag{7.3}$$

The element ϕ is a damping parameter on the trend. It is designed for the flattening of the trend in the long run. This is so that the short-term trend b_t converges on the long-term trend value b, rather than on zero. The element α is the smoothing parameter on the ARMA error component.

Trend Equation

The trend equation is given by:

$$b_t = (1 - \phi)b + \phi b_{t-1} + \beta d_t \tag{7.4}$$

The element b captures the long-term trend in the data. Ignoring βd_t for the moment, the current trend (b_t) is a weighted combination of the long-term trend (b) and the short term trend (b_{t-1}) with the weight determined by the damping parameter ϕ.

Now, β is a smoothing parameter on the ARMA error component. So, the long-term trend is a weighted combination of the short and long term trends plus a smoothed error component.

Seasonality

The equation defining seasonality is given by:

$$s_t^{(i)} = s_{t-m_i}^{(i)} + \gamma_i d_t \tag{7.5}$$

where similar to equation 7.3 and equation 7.4,γ_i is a smoothing parameter on the ARMA error component.

Error component

There is a single source of error that drives the level, trend and seasonal components. It is defined as an ARMA (p,q) model:

$$d_t = \sum_{i=1}^{p} \varphi_i d_{t-i} + \sum_{i=1}^{q} \theta_i \varepsilon_{t-i} + \varepsilon_t \tag{7.6}$$

where ε_t is the error processes assumed to be normally distributed with a mean of zero and constant variance equal to σ^2.

Special cases

Equation 7.1, equation 7.2, equation 7.3, equation 7.4, equation 7.5, and equation 7.6, are collectively called the BATS model as an acronym for the key features of the model: <u>B</u>ox and Cox transformation, <u>A</u>RMA errors, <u>T</u>rend and <u>S</u>easonal components.

The BATS model is often denoted by the shorthand BATS$(\omega, (p, q), \phi, m_1, ...m_1, m_2, ..., m_T)$. This is useful because it turns out that the BATS$(1, (0, 0), 1, , m_1)$ model corresponds the Holt-Winters model we discussed on page 47.

TBATS Model

The TBATS model is similar to the BATS model. The main difference between the two models is the introduction of a trigonometric representation of the seasonal components (hence

TBATS). Each seasonal pattern is represented by a specific Fourier series form where:

$$s_t^{(i)} = \sum_{j=1}^{k_i} s_{j,t}^{(i)} \tag{7.7}$$

Where k_i is the number of harmonics required for the ith seasonal component calculated as:

$$k_i = \begin{cases} \frac{m_i}{2}, & for\ even\ m_i \\ \frac{m_i-1}{2}, & otherwise \end{cases} \tag{7.8}$$

The stochastic level of the ith seasonal component $(s_{j,t}^{(i)})$ is modeled by:

$$s_{j,t}^{(i)} = s_{j,t-1}^{(i)} \cos\left(\frac{2\pi j}{m_i}\right) + s_{j,t-1}^{*(i)} \sin\left(\frac{2\pi j}{m_i}\right) + \gamma_1^{(i)} d_t \tag{7.9}$$

And the stochastic growth in the level of the ith seasonal component required to model the change in the seasonal component by time $s_{j,t}^{*(i)}$:

$$s_{j,t}^{*(i)} = -s_{j,t-1}^{(i)} \sin\left(\frac{2\pi j}{m_i}\right) + s_{j,t-1}^{*(i)} \cos\left(\frac{2\pi j}{m_i}\right) + \gamma_2^{(i)} d_t \tag{7.10}$$

The coefficients $\gamma_1^{(i)}$ and $\gamma_2^{(i)}$ are smoothing parameters, d_t the ARMA(p,q) error process.

TBATS uses a trigonometric function. This allows the estimation of non-integer seasonal frequency. This might be useful for daily observations if you want to define a calendar year as 365.25 days or 52.18 weeks per year to take into account leap years.

> ***NOTE...*** ✍
>
> The seasonal components become deterministic by setting the smoothing parameters $\gamma_1^{(i)}$ and $\gamma_2^{(i)}$ equal to zero.

Interpreting Mean Absolute Scaled Error

The Mean Absolute Scaled Error (MASE) is a performance metric used to compare performance between different series:

$$MASE = \frac{\sum |y_t - \hat{y}_t|}{\sum |y_t - y_{t-1}|}$$

The denominator is essentially the cumulative forecast of a random walk. Hence, MASE is a measure of relative forecast accuracy relative to a naive model.

MASE is independent of the scale of the data, this allows it to be used to compare predictive accuracy across different data sets that are measured on different scales.

The key thing to remember when using this method is that MASE $>$ 1 indicates that in-sample one-step forecasts (from a naive forecast) perform better than the model under consideration. In other words, the naive forecast performs better than our estimated model. Ideally, we want MASE $<$1, where the estimated model performs better than a naive model (random walk).

> ***NOTE...*** ✍
>
> The in-sample MAE is used as the denominator because it effectively scales the errors.

Advantages of BATS and TBATS

The key feature of BATS and TBATS models is that they allow you to model a wide variety of seasonal patterns.

The TBATS model is useful when a time series with a complicated seasonal pattern is encountered. It can deal with complex types of seasonality such as multiple seasonal components. It also handles nonlinear features in time series. Finally, both BATS and TBATS take autocorrelation in residuals into account.

The TBATS model reduces the parameter space without a negative impact on the predictive accuracy. It requires the estimation of $2\,(k_1 + k_2 + k_3 + ... + k_T)$ initial values. This number is generally smaller than the number of seed parameters required by BATS models.

Example - Forecasting Cardiovascular Mortality

In the *Clever Wife*, a short story in *Tiger by the Tail*, Lindy Soon Curry elegantly captures the heart of traditional Korea. A young man on an urgent journey across that beautiful country comes face to face with a tiger, enough to give an ordinary person a heart attack!

> "*The poor man's heart sank for he realized that he had lost his mother's medicine and soon his own life. When he thought of his mother and how she had cared for him as a child, he suddenly got a rush of energy, and he clung to the tiger's tail. But his energy gradually diminished, his hand were blistered, and his legs buckled under him. Finally, as the tiger breathed into his face and then roared a triumphant roar, the man collapsed in a dead faint.*"

Today the capital of South Korea is Seoul, an international metropolis. The tigers reside (mainly) in the zoo, and tourists can move around that great city quite easily, without the fear of a rapidly pounding heart, brought about by coming face-to-face with a tiger. In this section, we build BATS and TBATS models to forecast weekly levels of cardiovascular mortality in Seoul.

Step 1 – Collect, Explore and Prepare the Data

Daily levels of cardiovascular mortality for the city of Seoul can be found in the HEAT package. We load the package and extract the data into a zoo object called cardio:

```
require(HEAT)
data("mort")
Seoul <-  read6city(mort, 11)
dates <- seq(as.Date("2000-01-01"),
as.Date("2007-12-31"),
by = "1 day")
library(xts)
cardio <- zoo(Seoul$cardio,
dates)
```

Most of the above code should be familiar to you by now. The function read6city extracts a city-specific data set from the data frame mort. The code for Seoul is 11, hence we pass that value as the second argument to the read6city function.

There is always the possibility of missing values in a data set of daily observations. We saw earlier that the sapply can be used to report a count of elements coded as missing:

```
sapply(cardio,
function(x)sum(is.na(x)))
```

```
[1] 0
```

It reports zero missing values.

Creating a weekly time series

Our interest lies in modeling weekly levels of cardiovascular mortality. To achieve this we have to translate the daily data to weekly values. Fortunately, weekly series can be created from the daily data using the apply.weekly function from the xts package:

```
weekly.avg <- apply.weekly(cardio,
mean)
```

The R object `weekly.avg` contains the weekly averages of the daily data. The first few observations can be viewed via the `head` function:

```
round(head(weekly.avg),2)

2000-01-02 2000-01-09 2000-01-16
     32.00      34.57      34.00
2000-01-23 2000-01-30 2000-02-06
     27.43      26.00      26.29
```

So, the first week (1st January 2000) had an average value of 32, and the sixth week (6th February 2000) an average value of 26.29.

We can use a similar method to inspect the last few observations:

```
round(tail(weekly.avg),2)

2007-12-02 2007-12-09 2007-12-16
     26.29      27.43      28.86
2007-12-23 2007-12-30 2007-12-31
     25.57      26.57      29.00
```

Figure 7.1 presents a time series plot of the values contained in `weekly.avg`. The data appear to fluctuate around an average value of approximately 25, with a peak value of 36, and a low of 17.

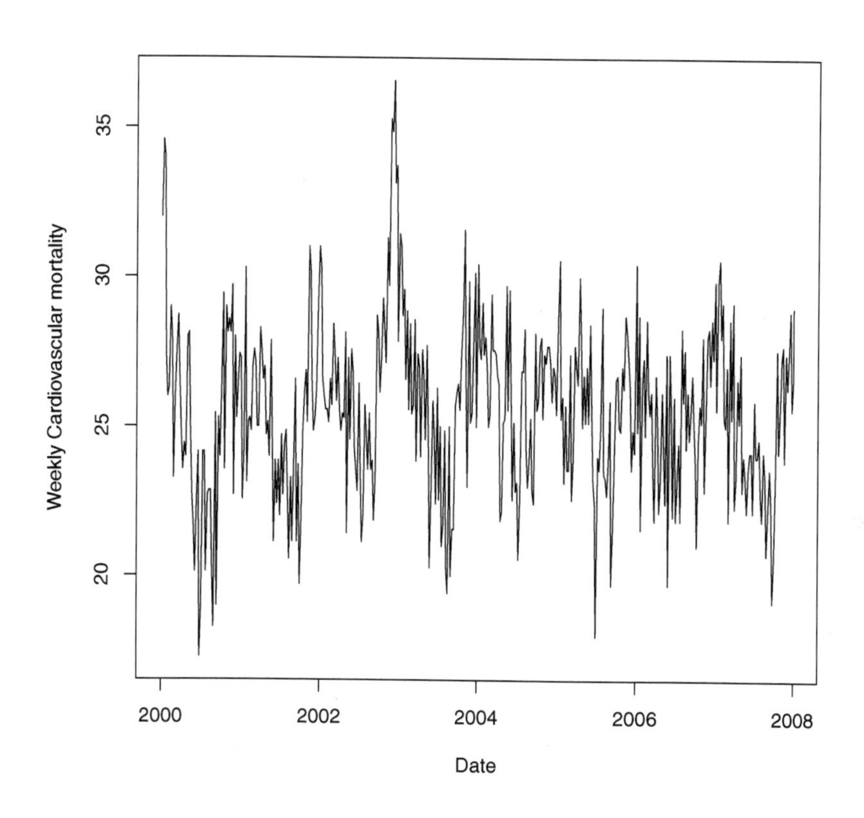

Figure 7.1: Weekly cardiovascular mortality

Using a train and test sample

For this illustration, we use a train and a test sample. The train sample is stored in the R object `data_train`, and the test sample in the R object `data_test`:

```
n=length(weekly.avg)
n_test=104
n_train=n-n_test
data_train<-ts(weekly.avg[1:n_train],
frequency=52, start=c(2000,01,31))
data_test<-ts(weekly.avg[(n_train+1):n],
```

```
frequency=52, end = c(2007,12,31))
```

The `n_test` argument is set to 104 to use the last two years of observations as the test set.

Figure 7.2 plots the train (top) and test samples (bottom). Both series fluctuate around a constant value of around 25.

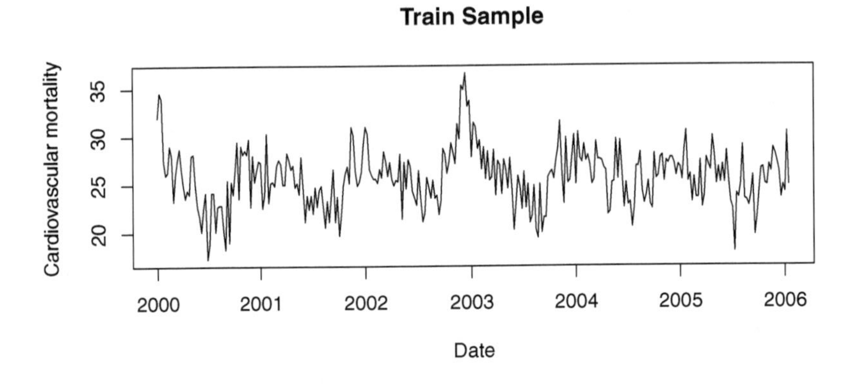

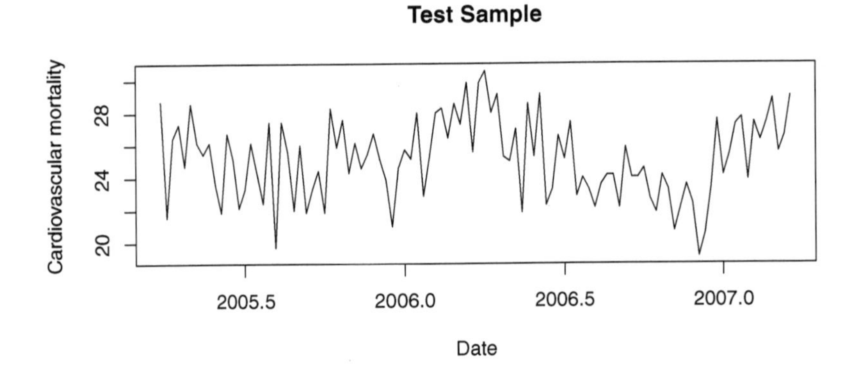

Figure 7.2: Train and test samples

Step 2 - Build a Forecast Model

Let's begin by building a BATS model. A simple way to do this is via the `bats` function in the `forecast` package. Simply pass the data to the function, it does all the work:

```
require(forecast)
fit_bats<-bats(data_train)
```

The fitted model is contained in the R object `fit_bats`:

```
fit_bats

BATS(1, {1,0}, -, {52})

Call: bats(y = data_train)

Parameters
  Alpha: 0.1176231
  Gamma Values: -0.2299629
  AR coefficients: 0.189648
```

Above we report the first few elements of output, and below we make a few observations on the output.

- It reports the optimal fitted model is a `BATS(1, {1,0}, -, {52})`. The first value reports ω in the Box-Cox transformation. Since it takes the value 1, the data were not transformed.
- The ARMA error component is reported next as `{1,0}`. For this data set the error is modeled as an ARMA(1,0) with AR coefficient of 0.189.
- We did not observe a trend in the data, and this is confirmed by the "-". If a trend were present it would contain the value of the damping parameter ϕ.
- The smoothing parameter α of the ARMA error component of the level equation is estimated as 0.117.
- Finally, one seasonality is estimated at 52 weeks, and the smoothing parameter on ARMA error component γ = -0.229.

You can view the BATS parameters directly by using $ to append the required item to the fitted model. For example, the following snippets of R code report the estimated values of α, γ and the AR coefficient:

```
round(fit_bats$alpha ,2)
```

```
[1] 0.12
```

```
round(fit_bats$gamma.values ,2)
```

```
[1] -0.23
```

```
round(fit_bats$ar.coefficients ,2)
```

```
[1] 0.19
```

Visualizing components

You can visualize the components of the fitted model using the `plot` function:

```
plot(fit_bats)
```

Figure 7.3 shows the resultant plot. The observed values are illustrated in the top image, followed by the level of the data. This is followed by seasonality, which exhibits an annual cycle. There is no trend in the data, hence this component is absent from the image.

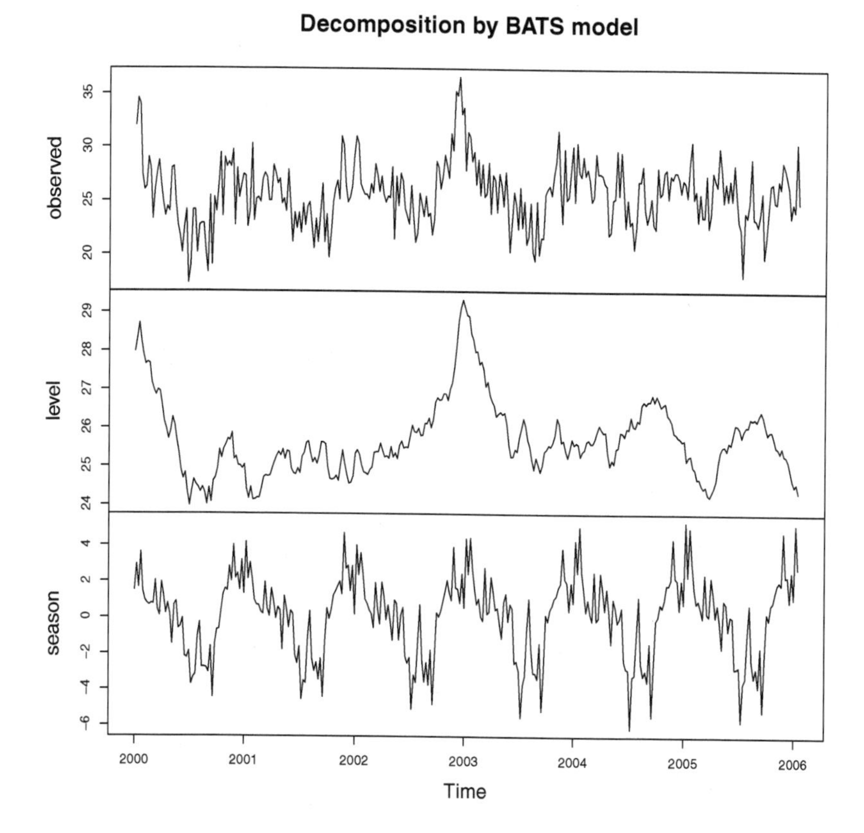

Figure 7.3: BATS model decomposition

Fine tuning BATS

You can control how the function `bats` fits the data by specifying additional arguments:

- To use the original data rather than a Box-Cox transformation add the argument `use.box.cox=FALSE`; The default is `NULL` which tries both with and without the Box-Cox transformation. The best fit is selected by AIC.

- To leave out the trend component set `use.trend=FALSE`; The default is `NULL` which tries both with and without a trend. The best fit is selected by AIC.

- If you do not want to include a damping parameter in the trend you can add the argument `damped.trend= FALSE`. The default is `TRUE`.
- You can also ask the function to fit a model without ARMA errors. This is achieved by adding the argument `use.arma.errors=FALSE`. The default is `TRUE` and the best fit is selected by AIC.

Step 3 - Evaluate Model Performance

You can perform an in-sample prediction using the forecast function. For example, here is how to forecast the last 12 weeks of the training set data alongside a 95% prediction interval:

```
pred<-forecast(fit_bats,
h=12,
level=0.95)
```

The above should be familiar to you by now. The parameter h controls the number of predictions, in this case 12 (weeks).
To visualize the data and prediction use the `plot` function:

```
plot(pred)
```

Figure 7.4 shows the original time series alongside the predicted values (in blue), the shaded region around the predictions represents the prediction interval.

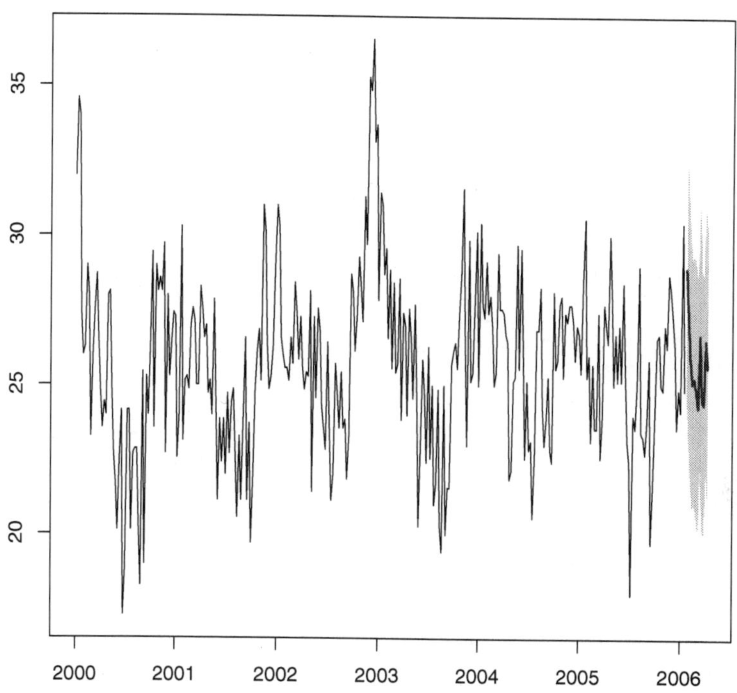

Figure 7.4: Predictions using the training set data

Assessing out of sample performance

Now, we are ready to assess model performance using the test sample. This is straightforward via the `bats` function. You simply pass it the test sample (`data_test`) and the fitted model (`fit_bats`):

```
f_bats <- bats(data_test,
model=fit_bats)
```

The R object `f_bats` contains the results.

So how well did the model perform? The `accuracy`

function from the **forecast** package returns several measures of forecast accuracy. We'll use it here to assess performance:

```
round(accuracy(f_bats),2)

                 ME RMSE  MAE   MPE MAPE
Training set -0.19 2.45 1.85 -1.36 7.61
             MASE  ACF1
Training set 0.75 -0.19
```

The function reports a number of performance metrics including the (ME), and MASE. The MASE is < 1 which indicates our model outperforms a naive forecast. However, MASE only really makes sense once you have multiple series on different levels to compare. For assessing accuracy on a single series, I prefer the MAE or RMSE.

Assessing the residuals

The output also reports the first value of the autocorrelation function of the residuals. In this example, it takes a value of -0.19. Recall ε_t, the error processes is assumed to be independently normally distributed with a mean of zero and constant variance equal to σ^2. Hence, we would expect (ideally) an autocorrelation of zero.

Figure 7.5 shows the auto correlations for the first ten lags of the residual. They are all within the dotted horizontal lines (95% confidence interval) and therefore not statistically significant from zero.

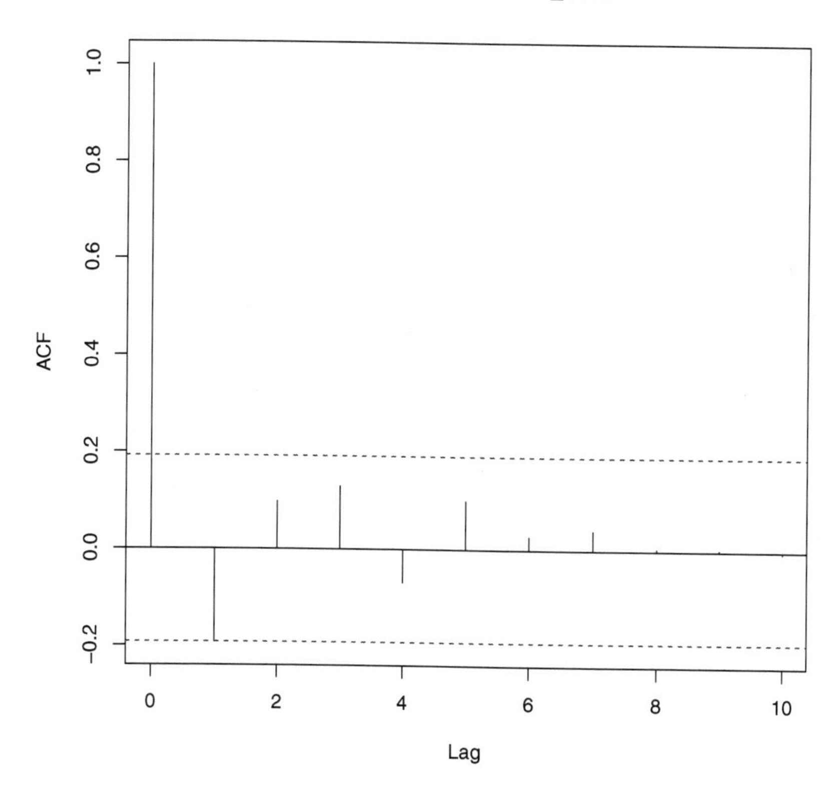

Figure 7.5: ACF of test sample residual

Figure 7.6 shows a time series plot of the residuals (top). They appear to fluctuate around a constant value, close to zero. You can see the exact value via the **mean** function:

```
round(mean(f_bats$error),2)
```

```
[1] -0.19
```

Not quite zero, we could assess it formally using a statistical test but we won't quibble here. In any-case, we can expect some random variation in the estimated value.

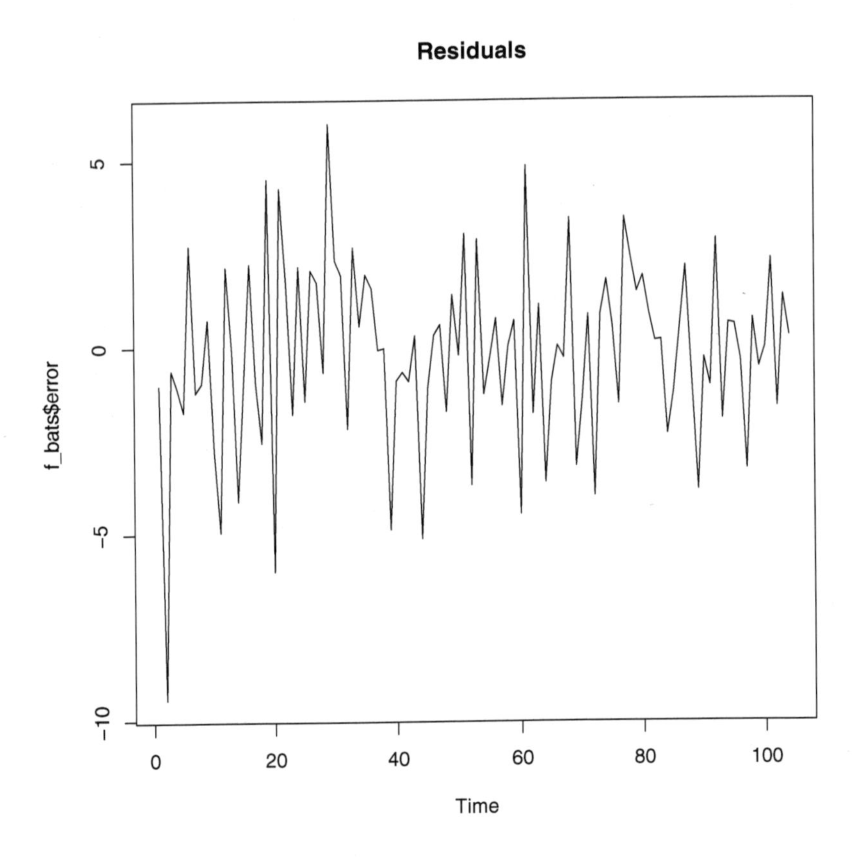

Figure 7.6: Time series plot of residuals

To assess the normality of the data we plot a density and QQ plot shown in Figure 7.7. The density plot is roughly bell shaped with an extended left tail; and for the most part the points in the QQ plot fall on the upward sloping line. However, there is some deviation in the left tail.

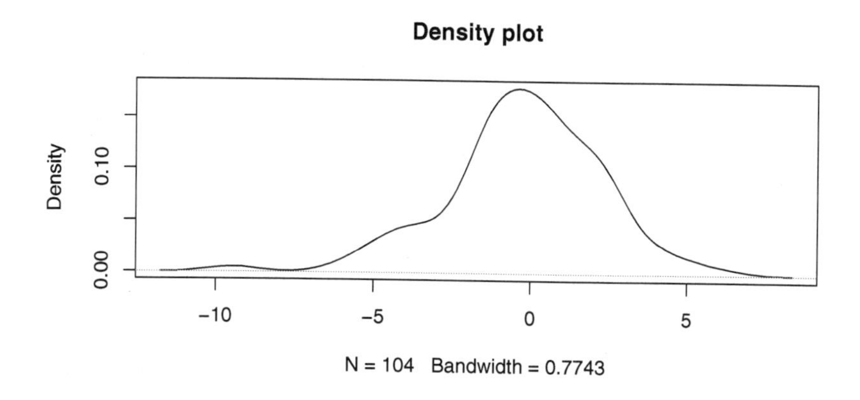

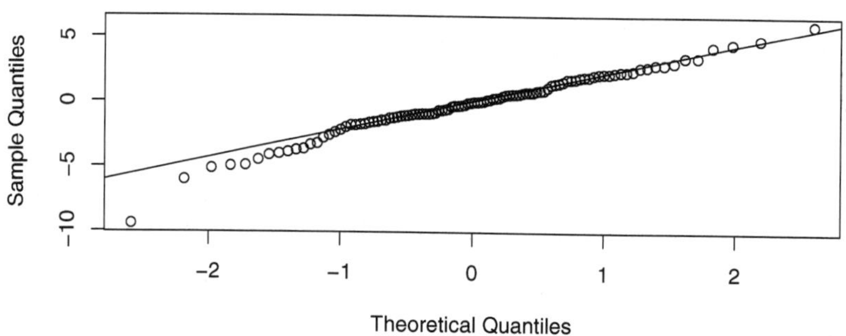

Figure 7.7: Density and QQ plot of residuals

Our initial observations on Figure 7.7 suggest we should investigate the assumption of normality formally via a statistical test. The Shapiro-Wilks normality test will do the job for us. It can be estimated via the **shapiro.test** function:

```
shapiro.test(f_bats$error)

        Shapiro-Wilk normality test

data:   f_bats$error
W = 0.97722, p-value = 0.06977
```

The null hypothesis of the Shapiro-Wilk normality test is that the residuals are from a normal distribution. The p-value of the test statistic, at 0.07, is greater than 0.05, and we cannot reject the null hypothesis that the residuals are normally distributed at the 5% level of statistical significance.

> ***NOTE...*** ✍
>
> Remember, we reject the null hypothesis if the probability (`p-value`) is "very small". Of course "very small" is determined by the level of statistical significance you select before running the test. In this example, we used a 5% level of statistical significance. Other common levels are 10%, 1% and 0.1%.

Step 4 - Improving Model Performance

Now, let's see if the TBATS model can improve on the performance of the BATS model. It can be estimated via the `tbats` function:

```
fit_tbats<-tbats(data_train)
```

The fitted model is contained in the R object `fit_tbats`. You can visualize the components of the fitted model using the `plot` function:

```
plot(fit_tbats)
```

Figure 7.8 shows the resultant plot. The observed values are illustrated in the top image, followed by the level of the data. Unlike the BATS model, a slope is estimated. It initially rises quickly, then levels out somewhat, before a sharp plunge toward the middle of the time series. It ends on a strong upward trend. The bottom diagram in Figure 7.8 visualizes seasonality, which exhibits a strong annual cycle.

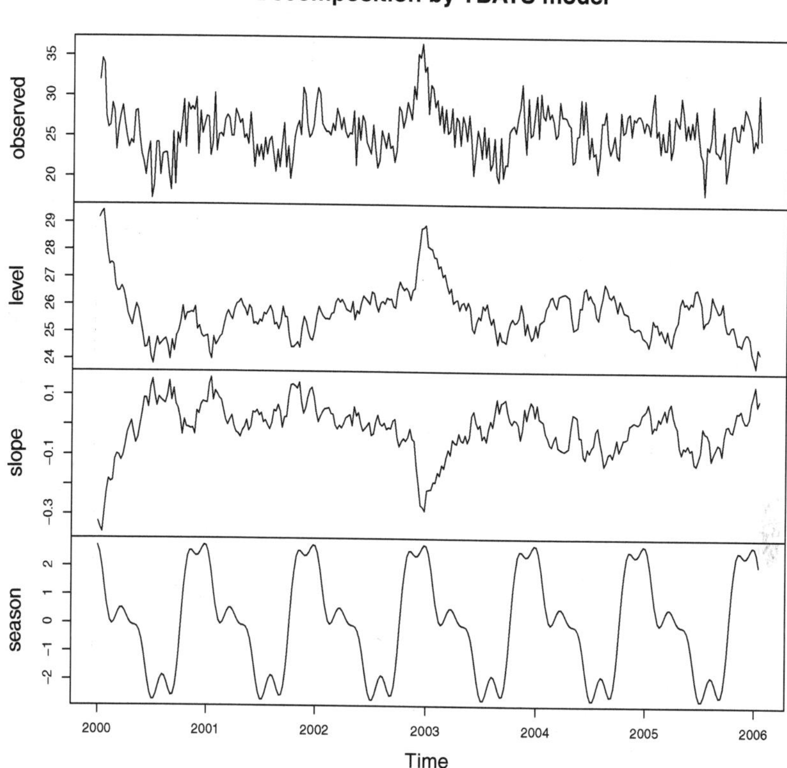

Figure 7.8: TBATS model decomposition

Estimated model and parameter values can be viewed by calling the fitted model:

```
fit_tbats

TBATS(1, {0,0}, 0.927, {<52,5>})

Call: tbats(y = data_train)

Parameters
  Alpha: 0.1433762
  Beta: -0.01368683
```

```
Damping Parameter: 0.926508
Gamma-1 Values: 6.172907e-05
Gamma-2 Values: 0.0001216391
```

Above we report the first few elements of output. Here are a few observations on this output

- It reports the optimal fitted model is a `BATS(1, {0,0}, 0.927, {<52,5>})`.
- The first value reports ω in the Box-Cox transformation. Since it takes the value 1, the data were not transformed.
- The ARMA error component is reported next as `{0,0}`. In other words, there is neither an auto regressive or moving average component.
- The smoothing parameter α of the error component of the level equation is estimated as 0.143.
- The smoothing parameter β of the error component of the trend equation is estimated as -0.013; and the damping parameter $\phi = 0.926$.
- Finally, two seasonal periods are estimated, the first at 52 weeks and the second at 5 weeks.

Following the procedure discussed earlier for the BATS model, we assess the performance of the `fit_tbats` on the test sample:

```
f_tbats <- tbats(data_test,
model=fit_tbats)

round(accuracy(f_tbats),2)

                ME RMSE  MAE   MPE
Training set -0.45 2.31 1.79 -2.45
             MAPE MASE ACF1
Training set 7.36 0.72    0
```

The TBATS model has a smaller RMSE (2.31 v 2.45) and MAE (1.79 v 1.85) than the BATS model. Overall, it appears to offer a marginal improvement over BATS.

NOTE... ✍

If you have knowledge on the nature of seasonality you can add it manually using the `msts` function via the argument `seasonal.periods`. Here is an example:

```
y <- msts(data_train,
seasonal.periods=c(7,52,12,365.25))
fit <- tbats(y)
round(accuracy(fit),1)
```

```
             ME RMSE MAE MPE MAPE
Training set  0  2.4 1.9  -1  7.7
             MASE ACF1
Training set  0.9    0
```

Limitations of BATS and TBATS

It is possible to fit ARIMA models with multiple seasonal components as Fourier terms, and often this approach will give similar results to the TBATS model. However, TBATS is preferable if seasonality changes over time, in addition the approach is fully automated. The core downside is that you cannot include explanatory covariates (such as time of day, or knowledge about an advertising campaign) as additional independent variables in the TBATS framework.

The BATS and TBATS models are based on a state space model with a single error. It is actually a generalization of the state space model underpinning the exponential smoothing models we discussed in chapter 4. These type of models are designed for use when there are multiple cyclic patterns (e.g. daily, weekly and yearly patterns).

Summary

The great thing about BATS and TBATS is that the modeling algorithm is entirely automated and does not require deep technical knowledge to apply it successfully. Simply pass it the data and the function does the work, It is that simple. That's not to say it is a simple or simplistic approach, in fact it is quite sophisticated as it allows for automatic Box-Cox transformation of the data and estimation of ARMA errors of the residuals. Both of these features were specifically included to enhance overall performance.

In the next chapter, we discuss another easy to use, highly accurate automated procedure, the multiple aggregation prediction algorithm.

Chapter 8

The Multiple Aggregation Prediction Algorithm

IN business it is common to convert a high frequency time series (say daily) into a lower frequency time-series (say quarterly). The Multiple Aggregation Prediction Algorithm (MAPA) builds on this idea. However, rather than selecting a single higher frequency series, MAPA generates several such series and uses them to generate a single forecast.

In this chapter, you will:

- Gain an understanding of how MAPA works.
- Work through the two steps involved in building a MAPA forecast.
- Understand the benefits of this approach for business organizations.
- Build several time series forecasts using MAPA in R.

The combination of forecasts produced from different frequencies of the same data allows a model to capture different aspects of the dynamics of our data.

Understanding the Multiple Aggregation Prediction

Multiple Aggregation Prediction Algorithm (MAPA) transforms time series data into several aggregates series. For example, a monthly time series might be transformed into bimonthly, quarterly and semi-annual series. Each series generates level, trend and seasonal estimates which are combined into an overall aggregate forecast.

Clarifying Temporal Aggregation

The process of non-overlapping temporally aggregating is outlined in Figure 8.1. It consists of two core steps.

Step 1: The original data is aggregated via a temporal aggregation procedure into a specific frequency of data. In the illustration, daily data is aggregated into weekly, monthly, bi-monthly, quarterly and annual data.

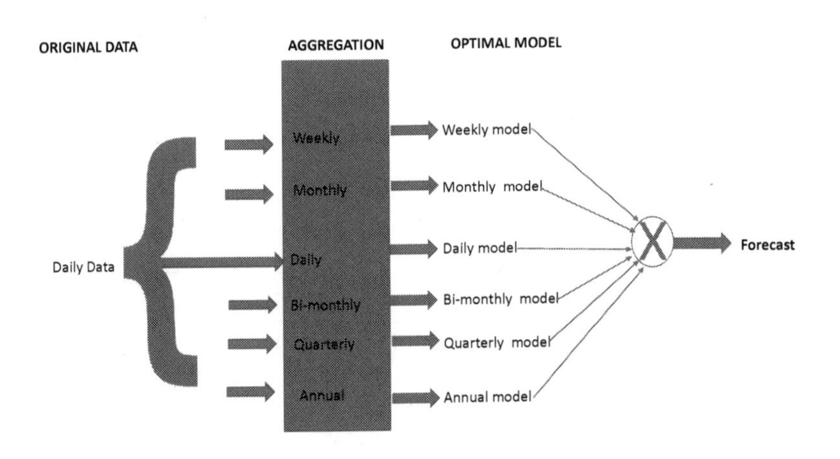

Figure 8.1: MAPA Overview

Multiple aggregation levels are created using non-overlapping means of length k. Given the original time series y_t , the kth temporally aggregated series is defined by :

$$y_i^k = \frac{\sum_{t=1+(i-1)k}^{ik} y_t}{k} \tag{8.1}$$

The increase of aggregation level results in time series of increasing lower frequency so that the aggregation step works as a filter, smoothing the high frequency features. This is because equation 8.1 acts as a type of moving average, so that the resulting time series is smoother than the original series.

For example, if you begin with a daily time series, high frequency components associated with day of the week are progressively filtered out as the aggregation level increases. This gradually dampens day of the week effects, seasonal impacts and the random component of a time series at each subsequent aggregation level. It also allows the lower frequency trend and level components to dominate at higher levels of aggregation.

NOTE... ✍

The average is generally used in equation 8.1 rather than the sum, because it maintains the original scale of the series across each aggregation level.

Step 2: A model is fitted to each aggregation level series, and their outputs in terms of level estimates, trend estimates and seasonal estimates, are combined to produce an aggregate forecast.

How many levels?

As a rule, you can continue transforming the original data series y_t to lower frequency aggregation levels as long as all transformed series have enough data to produce statistical forecasts. In practice, especially for business users, aggregation to a yearly frequency is often sufficient.

Which Forecasting Methods?

The MAPA framework can be used with any forecasting technique. In practice, it is often used with exponential smoothing models because these models separate a time series into level, trend and seasonal components during the modeling process.

For example, for each aggregation level series the appropriate exponential smoothing method is fitted. This can be carried out automatically using the component form exponential smoothing framework discussed in chapter 4. Next, the respective time series components are forecast. Then the time series components from each aggregation level are combined to construct the final forecast.

Advantages of Temporal Aggregation

When MAPA is brought to the attention of Managers and executives they are immediately attracted to the method. Part of the reason for this is the ease of understanding, but a more fundamental reason is it ensures the business analyst can produce forecasts that are reconciled for operational, tactical and strategic business horizons.

Consider the situation of a large commercial organization who had developed a massive spreadsheet based forecasting tool that generated predictions at the operational and strategic business horizons. For short-term operational forecasts weekly data was used; for tactical forecasts monthly data, and for strategic forecasts quarterly data.

For several years the forecasts produced from the different frequencies reconciled. Then one day they didn't. Unfortunately, the spreadsheet was broken, Little documentation, and the original creator of the spreadsheet had long left the organization. This is where MAPA shines - no spreadsheets, just simple to understand R code.

One of the other core benefits of MAPA is that characteristics inherent in the data at different aggregation levels are enhanced. This can help you better understand the dynamics of the time series, and it can also lead to substantial improvements in terms of forecasting performance. This is especially the case for longer horizons, as the various long term components of the series are better captured.

> ***NOTE...*** ✍
>
> Temporal aggregation at different levels, gives you multiple views of the data. This can be useful when trying to understand the dynamics of a time series.

Example - Forecasting Electricity Consumption

L. Frank Baum's 1901 story *The Master Key* ignited a passion for science in many children of that generation, in particular a fascination with electricity.

> *When Rob became interested in electricity his clear-headed father considered the boy's fancy to be instructive as well as amusing; so he heartily encouraged his son, and Rob never lacked batteries, motors or supplies of any sort that his experiments might require.*
>
> *"Electricity," said the old gentleman, sagely, "is destined to become the motive power of the world. The future advance of civilization will be along electrical lines. Our boy may become a great inventor and astonish the world with his wonderful creations."*

In this section, we use the MAPA framework to build models to forecast electricity consumption.

Step 1 – Collect, Explore and Prepare the Data

The sample is contained in the R package caschrono:

```
data("khct", package = "caschrono")
```

The R object khct contains monthly electricity consumption, heating degree days and cooling degree days for the period 1970-1984.

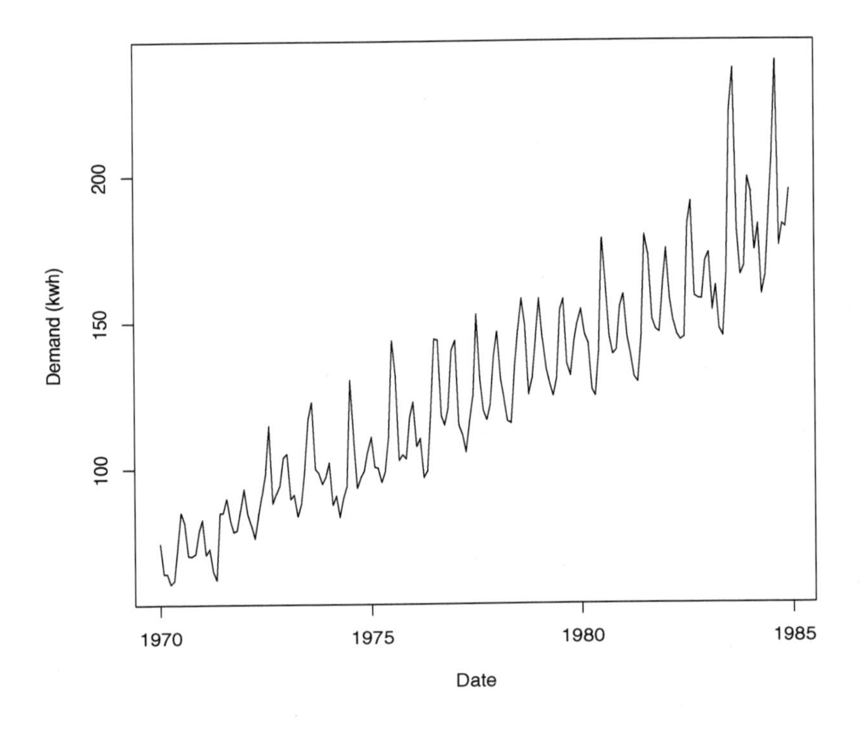

Figure 8.2: Electricity consumption

Figure 8.2 shows the time series plot of monthly electricity consumption (the series kwh in the data frame khct). It exhibits a strong upward trend with seasonality.

Step 2 - Build a Forecast Model

We build a model to forecast 24 months ahead. The object kwh contains measurement of electricity consumption in kilo-watt-hours, and for ease of illustration we transfer it to R object data_train:

```
n=length(khct[,"kwh"])
n_test=24
n_train=n-n_test
data_train<-ts(khct[1:n_train,"kwh"],
frequency=12,
start=c(1970,1))
```

A MAPA model can be estimated via the MAPA package and the mapaest function:

```
require(MAPA)
fit1<-mapaest(data_train,
type="ets")
```

The MAPA function uses the component form exponential smoothing framework. The type argument is used to select the exponential smoothing implementation to use. We set it to "ets" to use the forecast package. You can also use "es" to use the smooth package.

Details of the fitted model are stored in fit1:

```
fit1
```

```
MAPA fitted using ets    Original frequency: 12
Aggregation level: 1     Method: ETS(MAM)
Aggregation level: 2     Method: ETS(MAM)
Aggregation level: 3     Method: ETS(MAM)
Aggregation level: 4     Method: ETS(AAA)
Aggregation level: 5     Method: ETS(AAN)
```

```
Aggregation level: 6     Method: ETS(AAA)
Aggregation level: 7     Method: ETS(AAN)
Aggregation level: 8     Method: ETS(AAN)
Aggregation level: 9     Method: ETS(MAN)
Aggregation level: 10    Method: ETS(AAN)
Aggregation level: 11    Method: ETS(AAN)
Aggregation level: 12    Method: ETS(AAN)
```

The model fits twelve aggregation levels. The first level fits a **MAM**, and the twelfth level of aggregation was fitted using an **AAN** model. You can also use the plot function for a visual representation of the above:

```
plot(fit1)
```

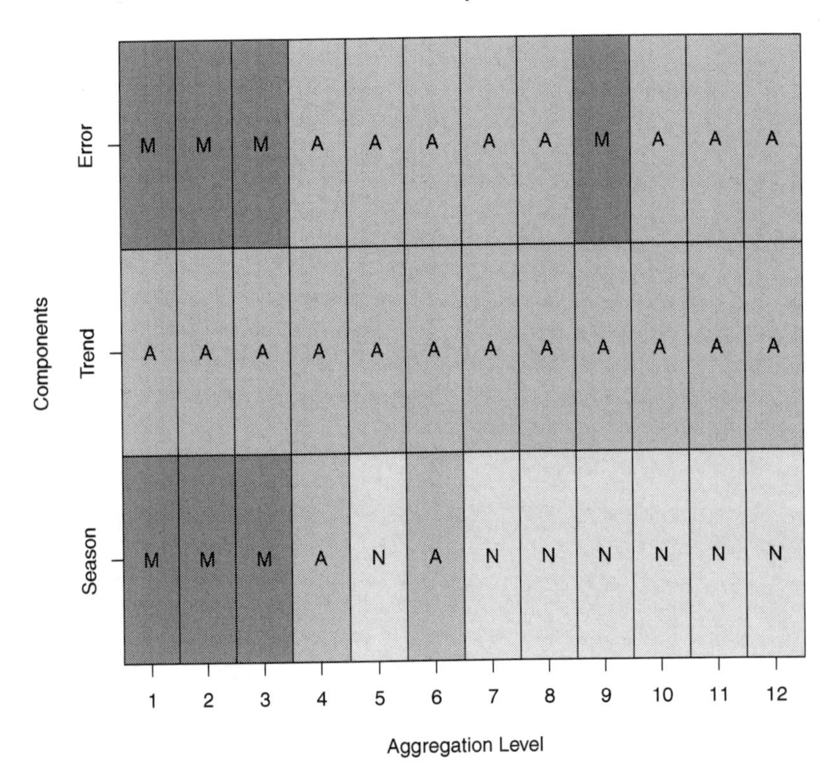

You can also view the level, trend and seasonality components for each level of aggregation. A simple way to visualize this is to use the **mapasimple** function:

```
fit1a<-mapasimple(data_train,
outplot=2,
type="ets",
fh=n_test)
```

The argument outplot=2 instructs R to print the output to the screen. You should see Figure 8.3. The top diagram presents the fitted values (blue), alongside the forecast (red). The bottom three diagrams provide a visual representation of the level, trend and seasonality by aggregation level.

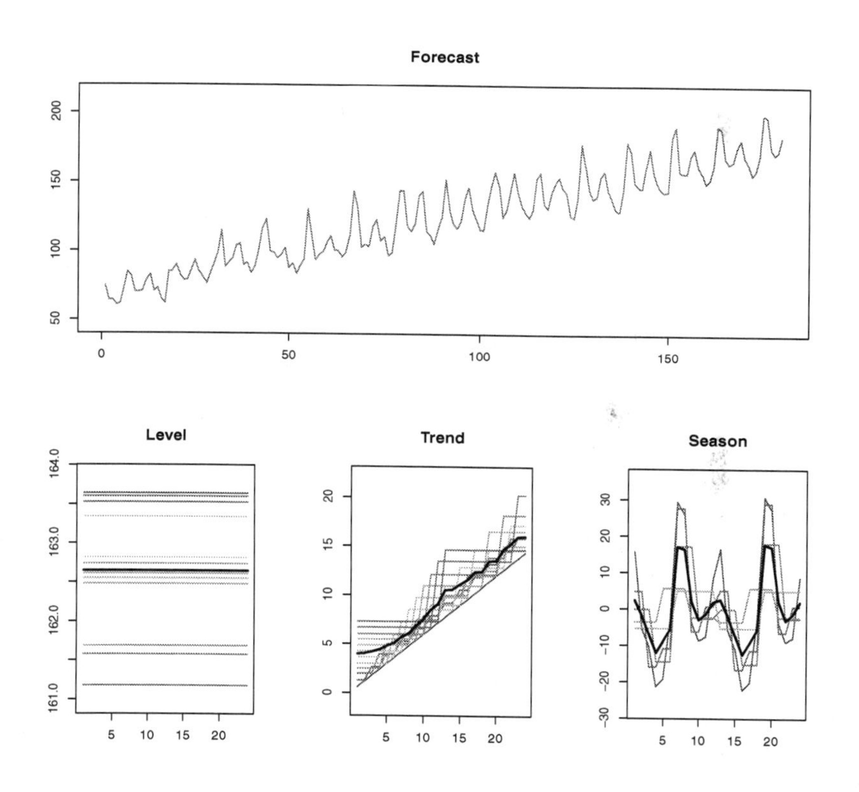

Figure 8.3: Trend, level and seasonal components by aggregation level

You can view individual aggregations by using the function `tsaggr`. For example, to see the third level of aggregation you would use:

```
tsaggr(data_train,fout=3,
fmean=TRUE,
outplot=TRUE)
```

Figure 8.4 shows the resultant plot. This level of aggregation catches much of the fluctuation in the underlying series.

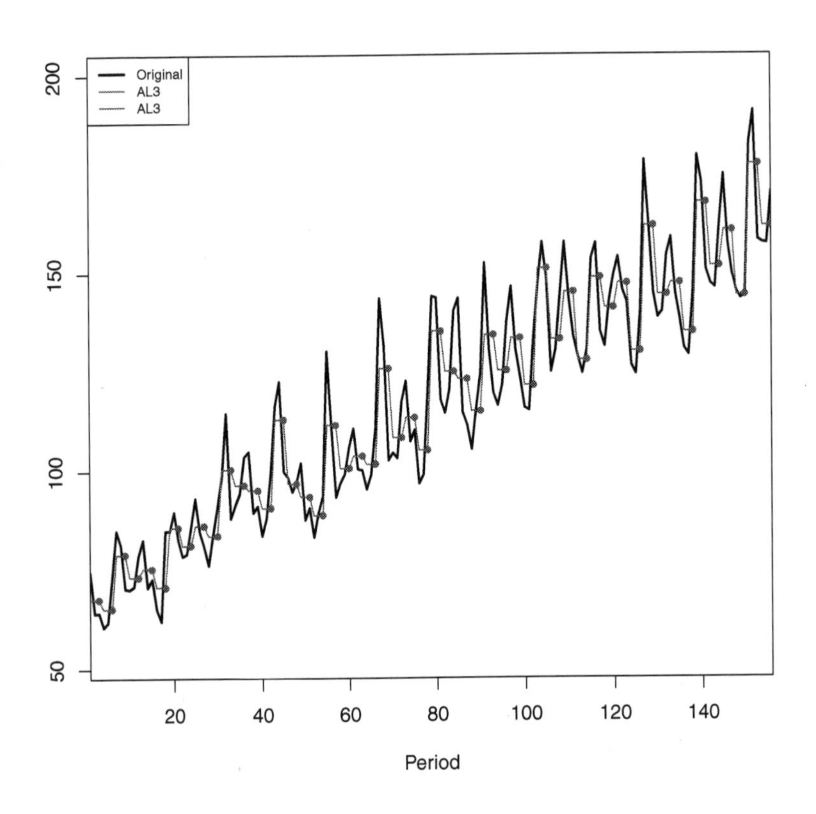

Figure 8.4: Setting `tsaggr = 3`

And to see the twelfth level of aggregation you would use:

```
tsaggr(data_train,
fout=12,
fmean=TRUE,
outplot=TRUE)
```

Figure 8.5 shows a resultant visualization. In this case the aggregation catches much of the longer-term dynamics of the data including the upward sloping trend. At this level, it's captured in the form of step to function.

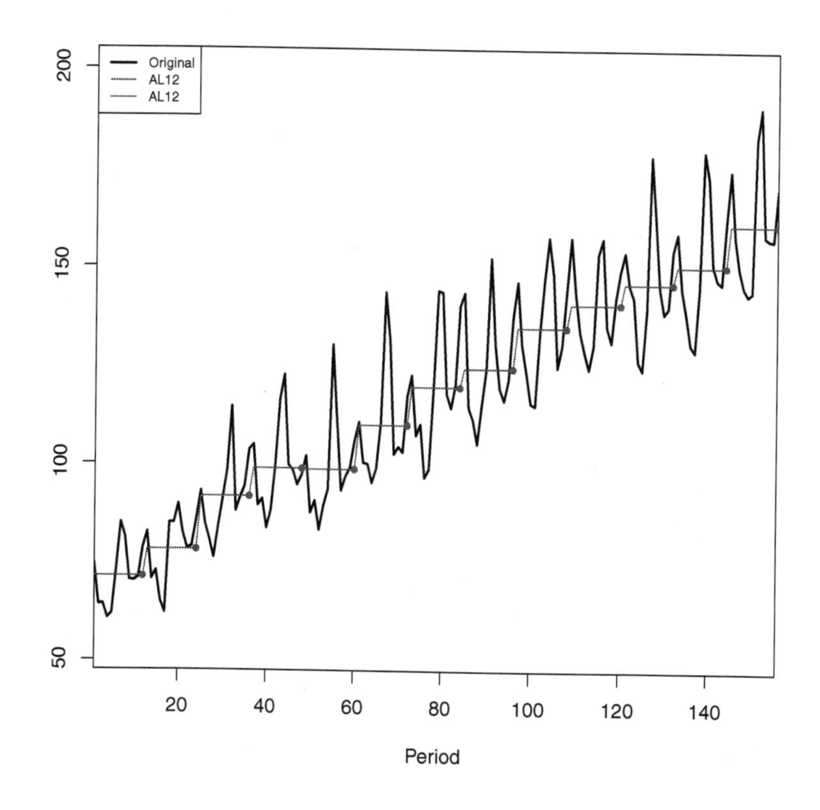

Figure 8.5: Setting `tsaggr = 1`

Step 3 - Evaluate Model Performance

Next, let's use the MAPA framework to generate forecasts. You can do this using the `mapafor` function:

```
fore1<-mapafor(data_train,
fit1,
fh=n_test,
conf.lvl=c(0.9,0.95,0.99))
```

We pass the `mapafor` function the sample data, the fitted model, the forecast, and the prediction intervals. For this illustration, we use a 90%, 95%, and 99% prediction interval. Let's take a look at the contents of the fitted object `fore1`. To begin we take a look at the actual predictions. We do this by appending the argument `outfor` for `fore1`:

```
round(fore1$outfor,1)
```

```
  t+1    t+2    t+3    t+4    t+5    t+6
174.5  162.1  157.4  149.8  152.9  163.4
  t+7    t+8    t+9   t+10   t+11   t+12
191.3  189.8  168.5  164.1  166.2  176.3
 t+13   t+14   t+15   t+16   t+17   t+18
181.8  168.8  164.1  156.3  159.7  170.5
 t+19   t+20   t+21   t+22   t+23   t+24
199.7  197.9  175.9  171.3  173.3  183.5
```

The first observation takes a value of 174.5, and the last observation takes a value of 183.5. We can also view the in-sample mean squared error by appending `MSE` to the fitted model:

```
round(fore1$MSE,1)
```

```
[1] 38.4
```

The mean squared error is reported as 38.4. You can use this as a benchmark against which to compare subsequent models.

NOTE... ✍

You can also view the mean absolute error by appending MAE to the fitted model.

An alternative technique

We can estimate the very same model as a fit1 using the mapa function. We save the fitted model in the R object fit1_alt. Here is how to do that:

```
fit1_alt<-mapa(data_train,
conf.lvl=c(0.9,0.95,0.99),
fh=n_test)
```

As a quick check, let's see if the model returns the same mean squared error:

```
round(fit1_alt$MSE,1)
```

```
[1] 38.4
```

Yep, we get precisely the same value. It turns out that the mapa function is a wrapper for the mapaest and mapafor functions.

Test set performance

Now, let's take a look at how the model performs with the test set data. We use the Metrics package, and it's MSE function to assess performance:

```
require(Metrics)
data_test<-khct[(n_train+1):n,"kwh"]
round(mse(data_test,fore1$outfor),2)
```

```
[1] 289.89
```

The test data is stored in the R object `data_test`. The out of sample mean squared error is reported as approximately 290, much higher than the in sample mean squared error. This is quite common with empirical models. The real issue is whether we can come up with an alternative model that has similar or better performance. We will take a look at this in the next section.

Step 4 - Improving Model Performance

One way to improve performance is to restrict the number of aggregation levels fitted in MAPA. This can be achieved by adding the `minimumAL` and `maximumAL` arguments to the `mapaest` function.

- The `minimumAL` argument controls the higher frequency elements;
- and `maximumAL` controls the lower frequency elements.

Another thing worth trying is using the `smooth` rather than the `forecast` the package.

Here is an example of how to do this where the fitted model is saved in the are R object `fit`:

```
fit<-mapaest(data_train,minimumAL=1,
maximumAL=4,
type="es")
```

Now let's use the `plot` function to visualize the fitted components by aggregation level:

```
plot(fit)
```

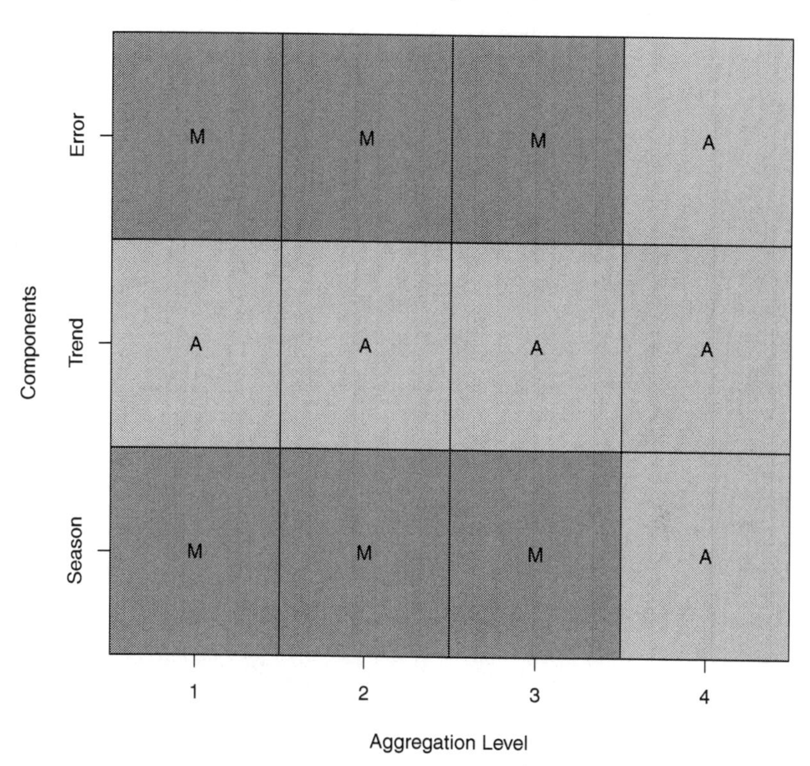

As expected the model contains four aggregation levels. The first three aggregation levels have a multiplicative error component, whilst the 4th aggregation level's error is additive. All four aggregation levels have an additive trend, while only the forth aggregation level has as an additive seasonal component.

"es" versus "ets"

One question that is frequently asked, is should we use "ets" or "es"? Unfortunately, there is no simple answer to this question. On some data sets you'll find the `forecast` package implementation will outperform, and on other samples the `smooth` package is the clear winner.

In many of the cases you will find the difference between the two is very small. To build successful time series forecasting models you have to experiment, and if time permits you should investigate both approaches to see which one performs best on your data.

Test set performance

As we did before, we use the mapafor function to forecast for a period of twenty-four months:

```
fore<-mapafor(data_train,fit,
fh=n_test,
conf.lvl=c(0.95))
```

Now take a look at the forecast values:

```
round(fore$outfor,1)
```

```
 t+1    t+2    t+3    t+4    t+5    t+6
174.9  161.4  157.2  148.7  152.6  164.0
  t+7    t+8    t+9   t+10   t+11   t+12
193.3  191.1  168.4  163.6  166.3  176.0
 t+13   t+14   t+15   t+16   t+17   t+18
183.0  168.8  164.4  155.5  159.5  171.3
 t+19   t+20   t+21   t+22   t+23   t+24
201.9  199.7  176.0  170.9  173.6  183.8
```

The first month prediction is 174.9, and the final prediction is 183.8. You'll notice that these values are very close to those predictions we saw in the previous section. Let's take a look at how well the model performs in terms of the mean square error:

```
round(mse(data_test,
fore$outfor),1)
```

```
[1] 269.9
```

The mean squared error for this model, at 269.9, is twenty points lower than that observed for the previous model. Nevertheless, it is significantly higher than the value observed for the training sample.

Figure 8.6 plots the predicted and observed values alongside a 95% prediction interval. As you can see the predicted values capture much of the dynamics of the observed values. However, the model under predicts extreme peaks in the data.

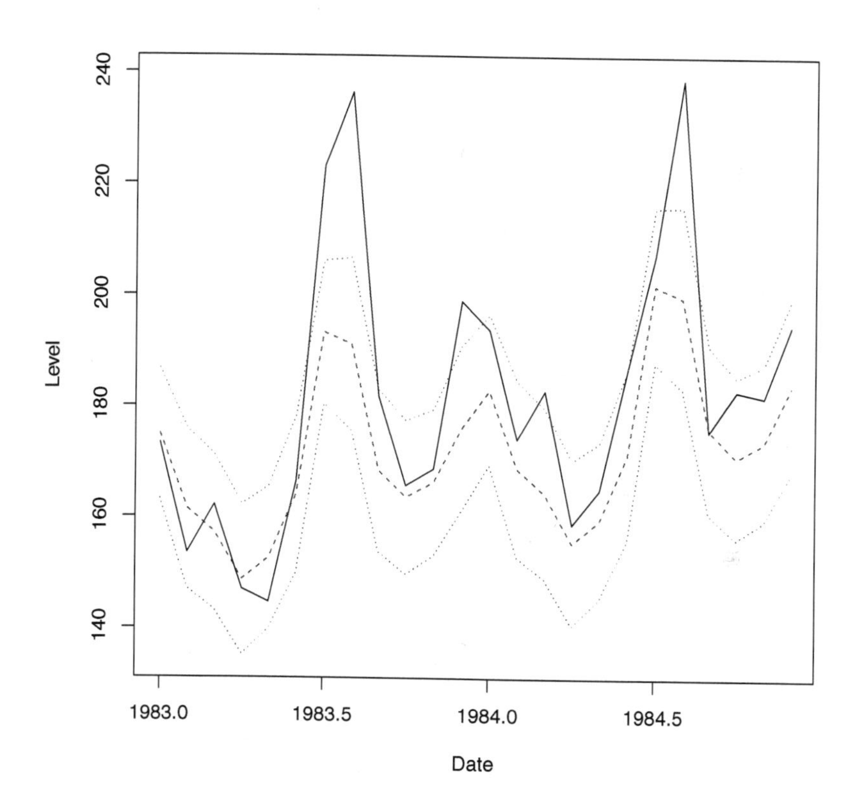

Figure 8.6: Actual and forecast values with 95% prediction interval

NOTE... ✍

The `mapafor`, `mapa` and `mapaest` functions have several more options, including setting the type of MAPA combination. Experiment with the options to see if you can enhance overall performance.

Limitations of Multiple Aggregation Prediction

Although MAPA can work with any forecasting tool, the decomposition into level and seasonality is a natural fit for exponential smoothing type models. The current implementations of the MAPA algorithm in R therefore use these methods. However, greater predictive flexibility and accuracy might be achieved by incorporating alternative predictive techniques, for example neural networks. At present, such techniques are not available within the `MAPA` package in R.

Summary

MAPA is a powerful framework for time series modeling. It is easy to intuitively grasp allowing transparency and direct implementation in business organizations. It captures both high and low frequency components, which leads to more accurate forecasts for both short and long horizons.

The idea is to take high-frequency data, say daily or weekly, generate lower frequency data, say quarterly or yearly, and then forecast each frequency combining them into an aggregate prediction. Its ability to combine forecasts across different levels of aggregation (daily, weekly, monthly and so on) is intuitively appealing. But the killer feature for business is that it leads to estimates that are reconciled across all the levels of aggregation.

It therefore provides consistent forecasts over the operational, tactical, and strategic, horizons.

The approach combines estimates from several models and therefore also diminishes the risk associated with the selection of a specific model.

In the next chapter, we explore another pragmatic approach to time series forecasting, a technique developed originally for use by Facebook, the Prophet algorithm.

Chapter 9

Effective Forecasting Using the Prophet Algorithm

When Facebook released the Prophet algorithm into the open-source community, it gave analysts another highly accurate pragmatic procedure for forecasting time series observations. Prophet was designed to be used without deep knowledge of the mathematical theories of forecasting or knowledge of advanced statistical techniques. In short, it was designed to be used by anyone, at any level of an organization, who has a need to generate forecasts.

In this chapter, you will:

- Gain an intuitive understanding of the components of a Prophet forecast.
- Identify the core advantages of this approach.
- Use the Prophet model in R to predict Wikipedia page views.

Although the Prophet algorithm is relatively new, its pragmatic approach and solid performance on business data make it an essential technique to add to your time series tool kit.

Understanding the Prophet Algorithm

The profit algorithm is a curve fitting additive regression model that decomposes a time series into four basic components. The components are growth, seasonality, holidays, plus an error term:

$$y_t = growth + seasonality + holidays + error \tag{9.1}$$

This looks similar to the additive decomposition we saw earlier in the text. However, there are some key differences. Let's take a look at each of the components.

Growth component

The growth component measures the growth in the underlying observations. It is modeled as a piece-wise linear or logistic growth curve trend. To give the model greater flexibility than is the case with a fixed trend model, the algorithm automatically detects changes in trend and adjusts the growth rate automatically.

Changes are modeled using coefficients that capture changes in trend at specific time points. These coefficients are called change-points, and the rate adjustment variable is modeled using a Laplace distribution with location parameter of 0.

As you can see from Figure 9.1, the Laplace distribution (also called the double exponential distribution) is uni-modal and symmetric. It is quite similar to the normal distribution except it has higher spike or peak, and slightly thicker tails.

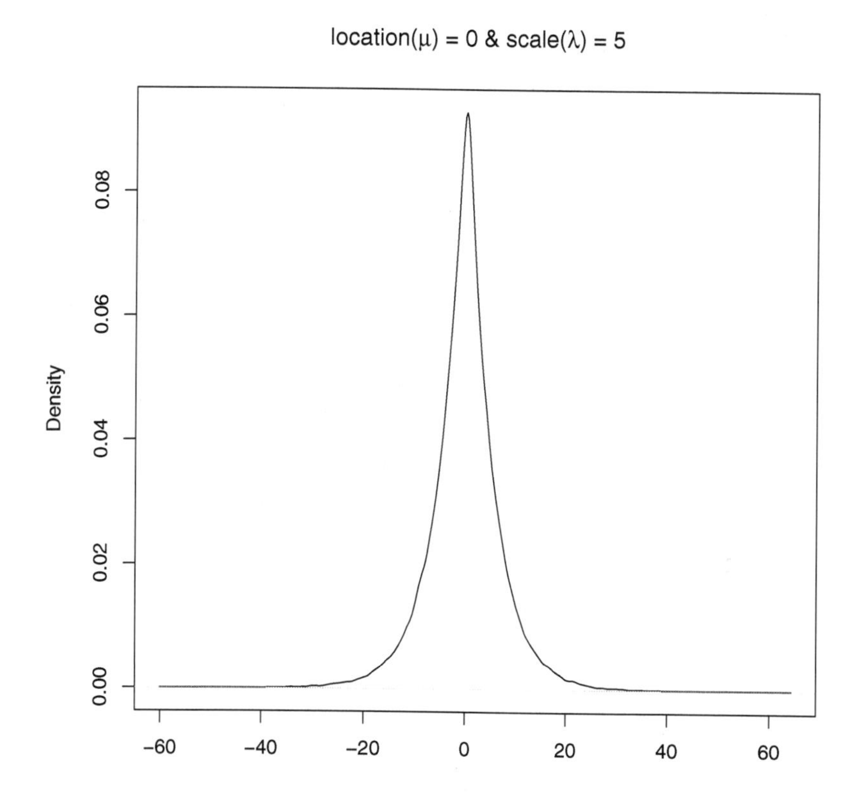

Figure 9.1: Laplace Distribution

Seasonality

Seasonality is the second major component of the Prophet algorithm decomposition. We have already observed that changes in the seasonal pattern of demand for goods and services is an important feature of many business time series. It is also important in numerous scientific time series, for example the level of many environmental variables fluctuates with the season.

The Prophet algorithm captures multiple levels of seasonality (day of week, day of year, etc.) modeled using Fourier series.

> ***NOTE...*** ✍
>
> For example, for weekly seasonality the number of approximation terms is 6, for annual seasonality it is 20. The seasonal components are then smoothed (using a normal prior).

Holidays

Holidays can have a significant impact on the dynamics of a time series. Across the course of a year, in any country, there are a number of major public holidays. Many of these do not occur on a specific date, but are determined by a complex formula, for example Easter. Others may celebrate a significant national event.

The algorithm makes use of a user supplied schedule to automatically adjust the dynamics of the model.

> ***NOTE...*** ✍
>
> For holidays, the model uses an indicator function that takes the value 1 on user supplied holidays. It then is multiplied by a smoothing factor (normal smoothing prior).

Error term

The error term is assumed to be normally distributed. It has a limited number of parameters so that analysts can make adjustments to the model based on their business needs.

NOTE... ✍

The Prophet algorithm is based on an additive model where non-linear trends are fit with yearly and weekly seasonality, plus holidays

Advantages of the Prophet Algorithm

The Prophet algorithm was designed to tackle a number of the biggest problems faced by businesses when it comes to time series forecasting. Many organizations have large volumes of time series data for which forecasting models are required. Manually building models when you have hundreds or thousands of time series which require prediction is inefficient and costly. The only practical solution is for the entire process to be automated.

There remains a considerable shortage in individuals highly trained in the statistical and mathematical theory that supports traditional time series forecasting techniques such as ARIMA. However, business organizations need forecasts and they are often produced by individuals who were not formally trained in statistics or predictive science. The Prophet algorithm was specifically designed to be used by individuals who do not necessarily have a deep and thorough knowledge of forecasting theory. It was also designed to be able to handle a wide variety of types of time series observations, and therefore offers a very flexible framework from which to generate predictions.

NOTE... ✍

The model is resistant to the effects of outliers, and supports data collected over an irregular time scale.

Example - Forecasting Wikipedia Page Views

Wikipedia is one of those websites that almost everyone who has been online has visited. It is hosted by the Wikimedia foundation and is the world's largest free online encyclopedia. The great thing about the Wikipedia model is that articles are created and edited by an army of volunteers scattered across almost every country in the globe.

In this section, we develop a Prophet model to forecast page views on the Wikipedia main page.

Step 1 – Collect, Explore and Prepare the Data

To build our forecasting model we need to extract the number of page views from the Wikipedia main page. There are many ways in which this can be achieved in R, perhaps the simplest is to use the `wikipediatrend` package. It provides convenient access to historical daily page view counts (Wikipedia article traffic statistics) stored at `http://stats.grok.se/`.

Downloading data from the internet

The `wp_trend` function will extract the information we need. The function takes three core arguments. The first, is the page you are interested in extracting page views from; we are interested in collecting data from the main page so we pass the argument "`main_page`" to the function.

The next set of arguments specify the date range over which you wish to collect the data; for this example, we will collect data from January 1, 2010 to January 20, 2016.

The final argument identifies the language for which you

want to collect the data, we choose English by setting the parameter lang equal "en":

```
require(wikipediatrend)
page_views <- wp_trend("main_page",
from = "2010-01-01",
to = "2016-01-20",
lang = "en")
```

The code will take a few moments to run as it extracts the historical data from the Internet. The R object page_views contains the downloaded data.

NOTE... ✍

The wikipediatrend R package relies onhttp://stats.grok.se/, which in turn relies on https://dumps.wikimedia.org/other/pagecounts-raw/ for the historical data. For page counts from 2015 on-wards you can visit https://dumps.wikimedia.org/other/pageviews/. This data is available in hourly intervals.

Viewing the observations

Let's use the head function to take a look at the first few observations:

```
head(page_views)
```

	date	count	lang	page	rank	month	title
1	2010-01-01	3970057	en	Main_page	2	201001	Main_page
2	2010-01-02	4210796	en	Main_page	2	201001	Main_page
3	2010-01-03	4067876	en	Main_page	2	201001	Main_page
4	2010-01-04	4285614	en	Main_page	2	201001	Main_page
5	2010-01-05	4568780	en	Main_page	2	201001	Main_page
6	2010-01-06	5058306	en	Main_page	2	201001	Main_page

We see, as expected, the first date is 1 January 2010, and under the `count` column for that particular day there were three million nine hundred and seventy thousand and fifty-seven page views. Under the `lang` column we see the `"en"`, this represents English.

We can use a similar procedure to look at the last few observations, using the `tail` function:

```
tail(page_views)
```

	date	count	lang	page	rank	month	title
1	2016-01-15	10672171	en	Main_page	2	201601	Main_page
2	2016-01-16	11381921	en	Main_page	2	201601	Main_page
3	2016-01-17	11714637	en	Main_page	2	201601	Main_page
4	2016-01-18	12198472	en	Main_page	2	201601	Main_page
5	2016-01-19	12449640	en	Main_page	2	201601	Main_page
6	2016-01-20	11823928	en	Main_page	2	201601	Main_page

As you can see, the final day reported is 20 January 2016. On this day there were eleven million eight hundred and twenty-three thousand nine hundred and twenty-eight page views of the main page in English.

Next, we create a data frame called `data_sample`. It contains the observations and their associated date:

```
y<-page_views[,"count"]/10000000
ds<-page_views[,"date"]
data_sample<-data.frame(ds,y)
```

Notice we have adjusted page views by 10,000,000, this is solely for presentation purposes.

Figure 9.2 shows a plot of the data. It is characterized by plateaus and periods of sharp upward growth alongside rapid declines. There are also giant peaks upwards which we speculate are related to major news events.

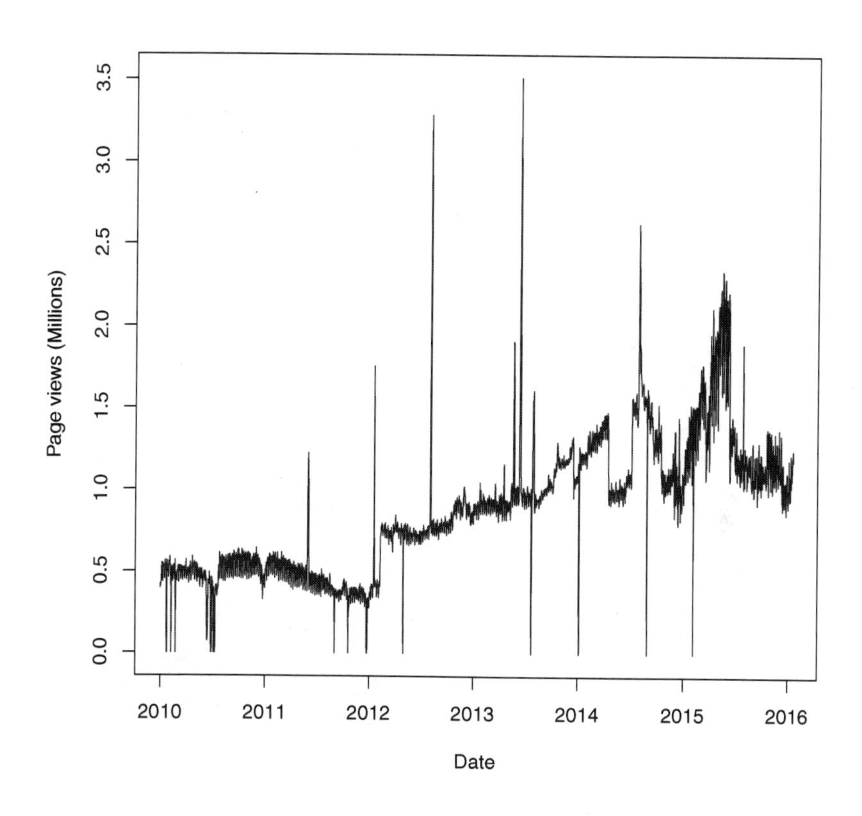

Figure 9.2: Plot of raw page view count data

Dealing with missing values

From Figure 9.2 you may have noticed a number of sudden downward spikes which appear to take the value zero, these are likely associated with missing values. Let's take a look to see how many observations behave like this:

```
data_sample$y[y == 0] <- NA
sum(is.na(data_sample$y))
```

```
[1] 20
```

The first line replaces the value zero with the character "NA"; then we use the sum function to count the number of "NA" observations. It reports 20 missing values. This is a relatively small number given that we have two thousand two hundred and ten observations in our sample. It turns out that the Prophet models are robust to outliers and missing observations. Nevertheless, we will investigate a little further.

One way to visualize the distribution of missing values is via a bar-plot. The plotNA.distributionBar function in the imputeTS package allows you to do this easily:

```
require(imputeTS)
plotNA.distributionBar(data_sample$y,
breaks = 100)
```

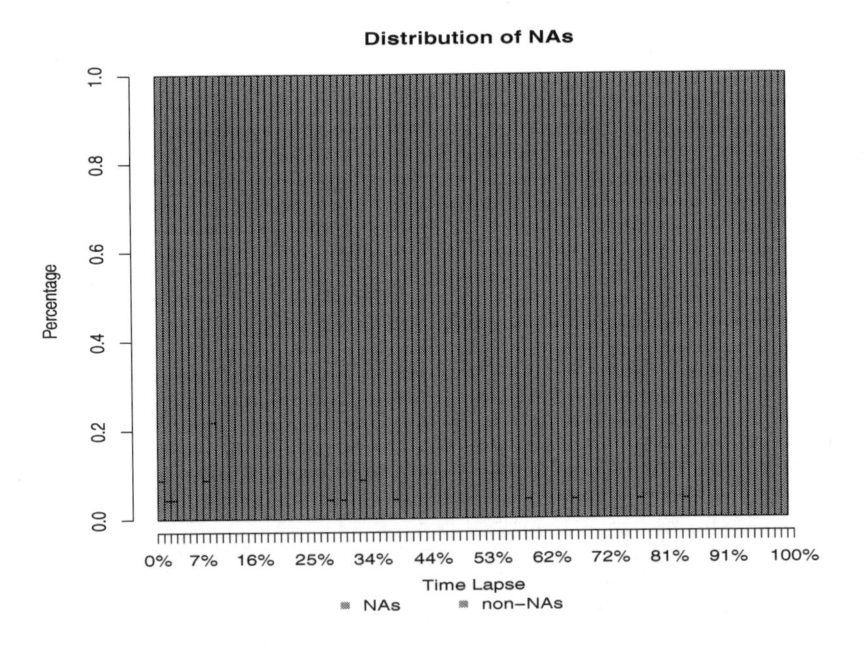

Figure 9.3: Bar-plot of distribution of missing values

Figure 9.3 conveys the overall impression that the number of missing values are rather small relative to the sample size; and

this confirms our prior finding. You can see from the diagram that the missing observations are distributed randomly across time (horizontal axis).
An alternative visual representation can be viewed using the `plotNA.distribution` function: again, as you can see from Figure 9.4, the missing values appear to be randomly distributed across time, although in some instances they cluster together.

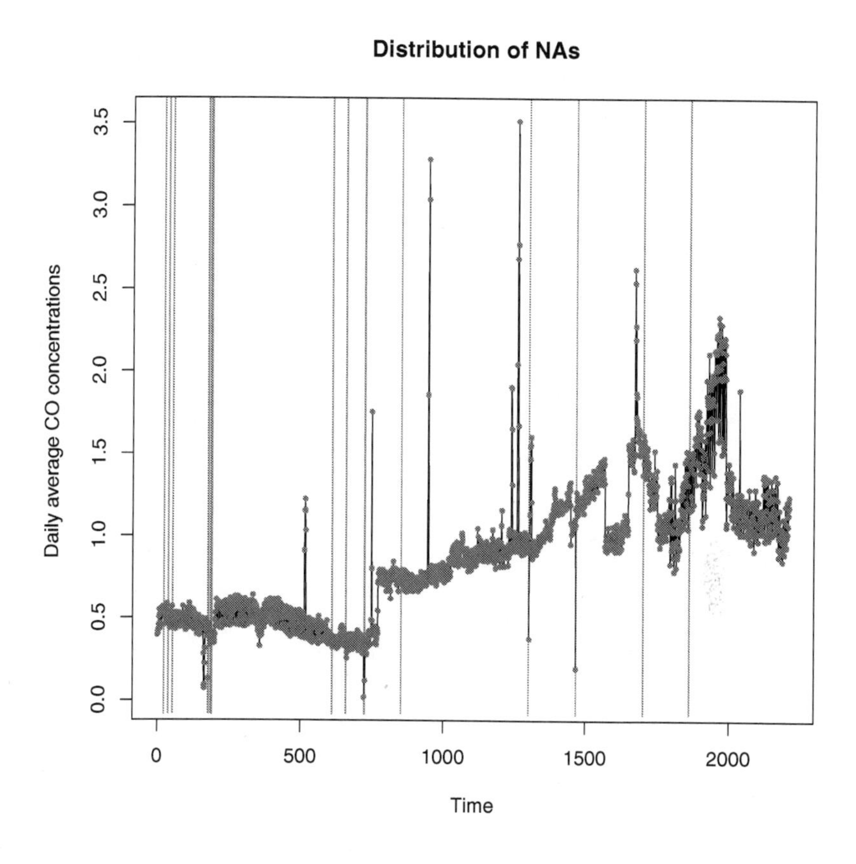

Figure 9.4: Time series plot of observations and missing values

As we mentioned earlier, the Prophet algorithm is robust to missing observations and outliers, so we could simply ignore these missing values given that they are a tiny proportion of the overall observations. Depending on the nature of the modeling task at hand this may or may not be acceptable. For illustrative purposes, we will use linear interpolation to replace the missing values with interpolated values. This can be achieved using the `na.interpolation` function from the `imputeTS` package:

```
data_sample$y <- na.interpolation(
  data_sample$y,"linear")
```

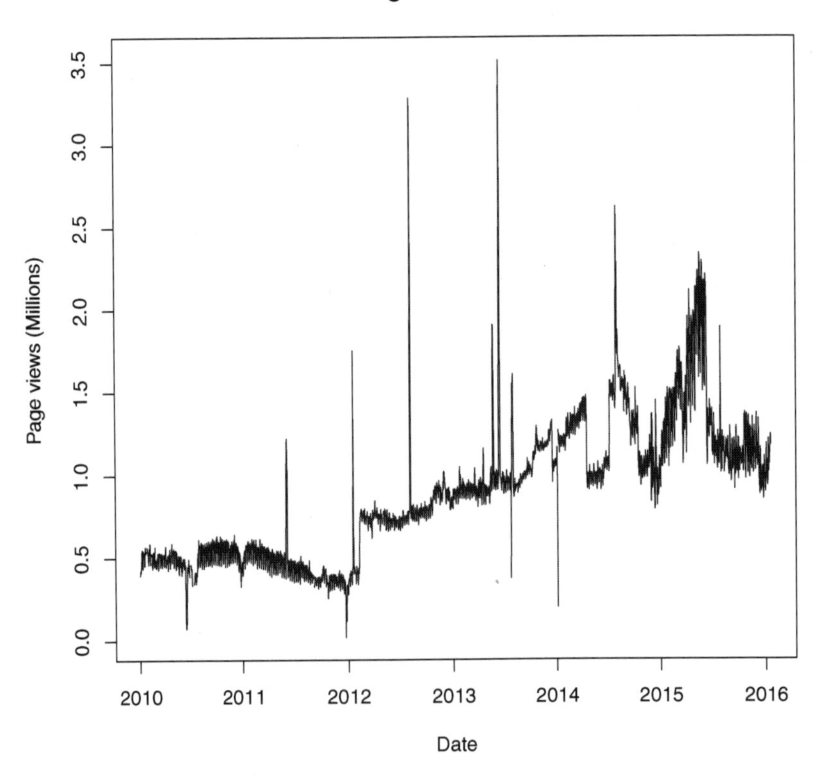

Figure 9.5: Cleaned page views data

Figure 9.5 shows the cleaned-up data. There are a few

things to notice about this clean time series; it retains the general dynamics of the original data, and there remain days where there are significant spikes upwards in page views, and days where there are significant downward spikes in page views.

Test and training sample

Now that we have the data in a suitable format, we'll need to create a train and test set. We use the `lubridate` and `dplyr` libraries to help us with this task:

```
require(lubridate)
library(dplyr)
day<-"2016-01-01"
data_train<-data_sample %>%filter(ymd(ds) <
    ymd(day))
data_test<-data_sample %>%filter(ymd(ds) >=
    ymd(day))
```

The R object `day` contains the cutoff date for the test and train samples. In this case 1 January 2016 and beyond are used as the test sample. Earlier observations are used to train the model.

The training sample is stored in the object `data_train`, and the test sample stored in the object `data_test`.

It is always useful to take a look at the observations in both the training and test set; we do this using the `head` and `tail` functions. First, the training set observations:

```
head(data_train)

          ds          y
1 2010-01-01 0.3970057
2 2010-01-02 0.4210796
3 2010-01-03 0.4067876
4 2010-01-04 0.4285614
5 2010-01-05 0.4568780
6 2010-01-06 0.5058306
```

Yep, it confirms the first observation occurs on 1 January 2010 and takes a value of approximately 0.397.

```
tail(data_train)
```

```
             ds         y
2185 2015-12-26 1.0453553
2186 2015-12-27 0.9656062
2187 2015-12-28 1.1095959
2188 2015-12-29 1.0072287
2189 2015-12-30 0.9981364
2190 2015-12-31 0.9372728
```

The final observation in the train set occurs in row 2190, which corresponds to 31 December 2015, with an observed value of approximately 0.937.

Let's do the same thing for the test sample observations:

```
head(data_test)
```

```
          ds         y
1 2016-01-01 0.8948959
2 2016-01-02 0.9848114
3 2016-01-03 0.9700719
4 2016-01-04 1.1880006
5 2016-01-05 1.0354326
6 2016-01-06 1.088472
```

As expected, the first date in the test sample is 1 January 2016, with an observed value of 0.894.

```
tail(data_test)
```

```
           ds        y
15 2016-01-15 1.067217
16 2016-01-16 1.138192
17 2016-01-17 1.171464
18 2016-01-18 1.219847
19 2016-01-19 1.244964
20 2016-01-20 1.182393
```

Our final observation occurs on 20 January 2016 (row 20), with an observed value of approximately 1.182.

Step 2 - Build a Forecast Model

Now that we have everything in place, let's go ahead and fit the model. The process is very simple and straightforward, you simply pass your training set data to the `prophet` function:

```
require(prophet)
set.seed(2019)
fit <- prophet(data_train)
```

As we have done before, the object `fit` contains the fitted model. As with all R objects you can view details of the characteristics using the `attributes` function:

```
attributes(fit)
```

```
$names
 [1] "growth"                   "changepoints"
 [3] "n.changepoints"           "yearly.seasonality"
 [5] "weekly.seasonality"       "holidays"
 [7] "seasonality.prior.scale" "changepoint.prior.
    scale"
 [9] "holidays.prior.scale"     "mcmc.samples"
[11] "interval.width"           "uncertainty.samples"
[13] "start"                    "y.scale"
[15] "t.scale"                  "changepoints.t"
[17] "stan.fit"                 "params"
[19] "history"                  "history.dates"

$class
[1] "prophet" "list"
```

Fitted and predicted values

Our real interest lies in using the model to predict future observations. In the `prophet` model framework, predicted and fitted values can be viewed by combining the make_future_dataframe function with the predict function as follows:

```
n_test<-nrow(data_test)
future <- make_future_dataframe(fit,
periods = n_test)
pred <- predict(fit, future)
```

Several observations are noteworthy about the above code.

- The object n_test contains a number of observations in the test set, in this case twenty;
- and the object future contains a combination of the test set dates and the training set dates;
- predictions are stored in the object pred, and the predict function takes the fitted model fit alongside the object future.

You can visualize the fitted and predicted values by passing the fitted model fit alongside the predictions in pred to the plot function as follows:

```
plot(fit, pred)
```

Figure 9.6 shows the resultant plot. The fitted and predicted values are denoted by the solid (blue) line, and the actual observations are denoted by the black dots. The chart also reports the uncertainty interval (shaded blue area) which is similar to a statistical prediction interval in the sense that it gives you a flavor of the degree or extent of uncertainty in the predicted values.

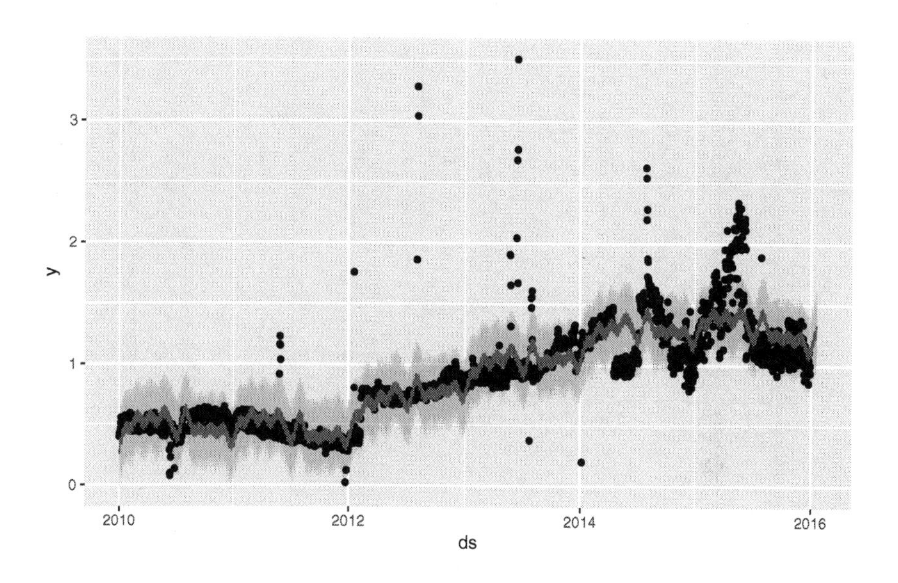

Figure 9.6: Predicted and fitted values for pages views

Step 3 - Evaluate Model Performance

From equation 9.1 we saw that the model decomposes observations into components including a trend or growth component and seasonality components. You can view these components visually using the `prophet_plot_components` function:

```
prophet_plot_components(fit,
pred)
```

You should now see Figure 9.7, it informs us the trend component was relatively flat between 2010 and the end of 2012 when it began to grow more steeply. It peeked towards end of 2014 then dipping slightly into early 2016.

There is also a marked day of the week effect which peaks on Monday and reaches a trough on Saturday. The seasonal effect at the yearly level appears highly smoothed (we can clearly see troughs in January and July).

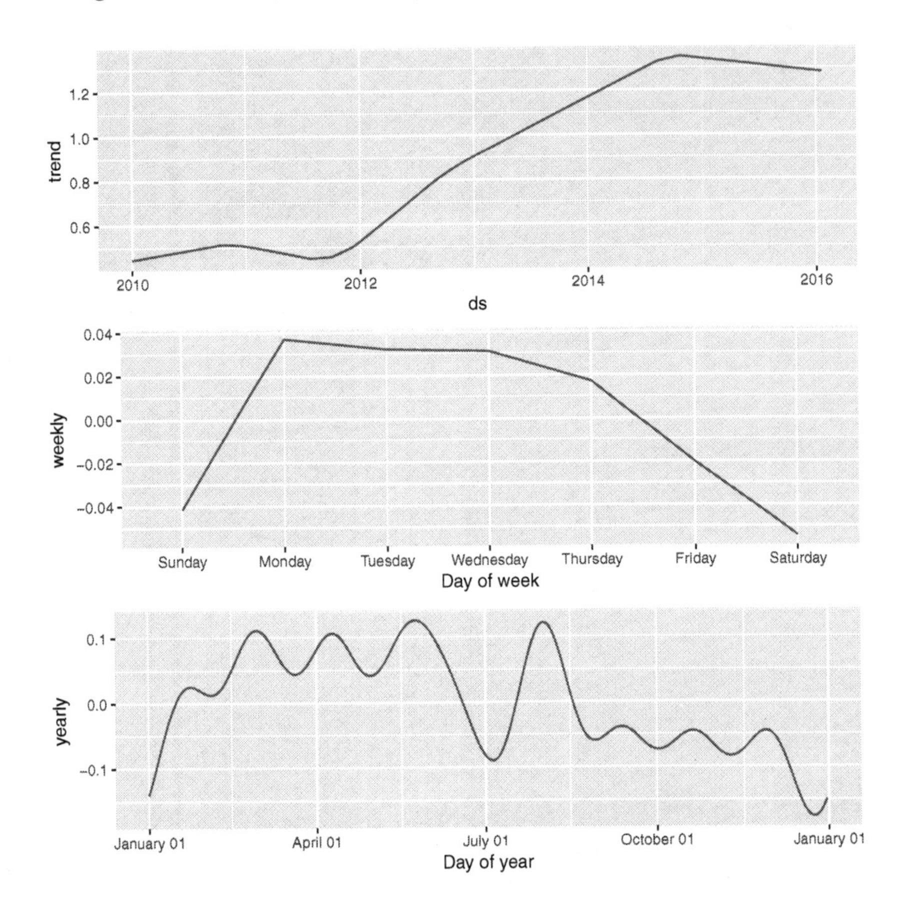

Figure 9.7: Components of fitted model

Test set performance

Now, let's take a look at how the model performed on the test set. First, we grab the test set predictions storing them in the object **pred_test**. We use the root mean squared error as our performance metric, loading it from the **Metrics** package:

```
pred_test<-pred %>%filter(ymd(ds) >= ymd(
   day))
require(Metrics)
round(rmse(pred_test$yhat,
data_test$y),4)
```

```
[1] 0.1615
```

The model produces a root mean squared error of around 0.16. We can use this value as a benchmark and against which to compare alternative models.

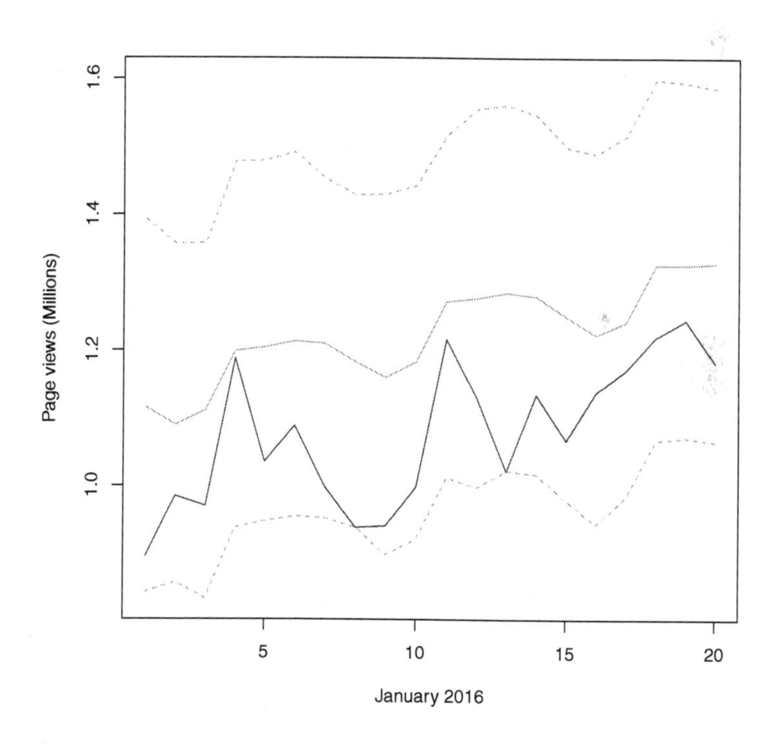

Figure 9.8: Predicted and actual values for page views

Figure 9.8 plots the actual observations (solid red line) against the predictions (solid gray line). As you can see, the actual observations are contained within the uncertainty interval, although the actual forecasts are systematically higher than the actual observed values. Nevertheless, the mean predictions (combined with the uncertainty interval) captures well the general dynamics of the observations.

Step 4 - Improving Model Performance

Our initial model captured much of the dynamics of real world data, this is one of the practical advantages of using the profit algorithm - it often works very well "straight out of the box." How can we improve the model further?

One technique that often works very well is to include country specific holidays into the modeling framework. Let's take a look at how we might do this.

Adding holidays

The first thing we need is a schedule of relevant holidays. We could manually create our own schedule, however for this illustration we'll use the **tis** package. It contains a neat little function called **holidaysBetween** that knows about federal (United States) holidays. You simply pass the function the date range of interest as the first two arguments. Here is how:

```
require(tis)
hols<-holidaysBetween(20100101,
20160120,
goodFriday = TRUE,
board = TRUE,
businessOnly = TRUE)
```

There are also several other options, we select Good Friday as a holiday along with the federal board vacation days, and also business vacation days are set to TRUE.

Let's take a look at the first few holidays for the year 2010:

```
head(hols)

     NewYears          MLKing   GWBirthday
     20100101        20100118     20100215
   GoodFriday        Memorial  Independence
     20100402        20100531     20100705
```

We see the holidaysBetween function has identified several key dates - New Year's, Day Martin Luther King Day, George Washington's birthday, Good Friday, Memorial day, and Independence Day.

Next, we create a data frame called df that contains the name of each holiday along with the date:

```
holiday<-names(hols)
ds<-as.Date(as.character(hols), "%Y%m%d")
df<-data.frame(holiday,ds)
```

Building the model

Now, we can use df as an additional input to the prophet function. There are also several hyper parameters that are interesting to play with. The changepoint_prior_scale argument often has a dramatic impact on performance. It is related to how strongly the model adjusts to trends. Large values allow many change-points, small values allow fewer change-points. The default value is 0.05, we increase it to 5:

```
set.seed(2019)
fit1 <- prophet(data_train,
growth="linear",
holidays=df,
changepoint.prior.scale=5)
```

Other parameters you might want to experiment with include:

- `n.changepoints`: Number of potential changepoints to include.
- `changepoints`: Dates of potential changepoints.
- `holidays.prior.scale`: Adjust the importance of holiday effects.
- `interval.width`: Sets the uncertainty interval, defaults to 80%.

Test set performances

Following the procedure we outlined earlier the predicted values are stored in the object `pred1`, and we use the `prophet_plot_components` function to visualize the components:

```
future1 <- make_future_dataframe(fit1,
periods = n_test)
pred1 <- predict(fit1, future1)
prophet_plot_components(fit1,
pred1)
```

Figure 9.9 displays each of the components. It is interesting to note that the trend exhibits an upward sloping series of humps then levels out towards the end of 2014. In addition, we see the plot also shows the impact of the holidays over the time period in question. The overall impact appears to be negative with many more bars below zero than above.

The pattern of weekly seasonality is similar to that we observed before, however the annual or yearly seasonality displays an upward trend towards June, and a downward trend thereafter.

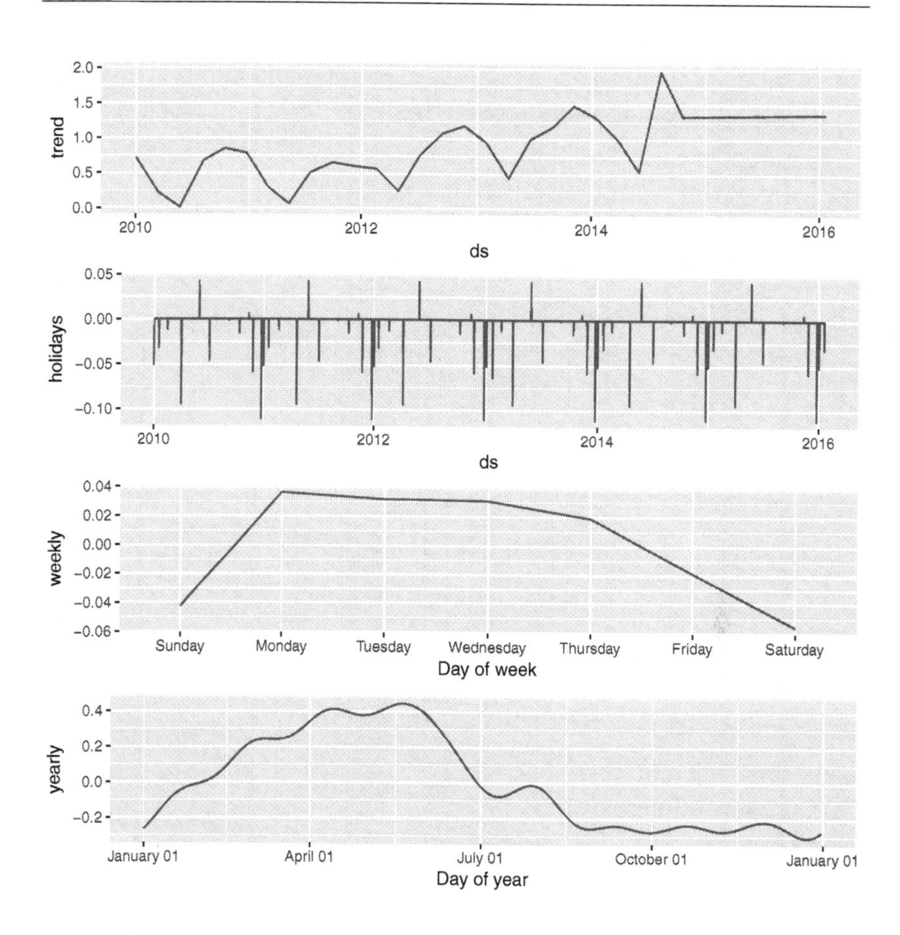

Figure 9.9: Components with holidays included

Root Mean Squared Error

Finally, let's take a look at the root mean squared error metric. We follow the exact same procedure as discussed earlier:

```
pred_test1<-pred1 %>%filter(ymd(ds) >= ymd(
   day))
require(Metrics)
round(rmse(pred_test1$yhat,
data_test$y),4)
```

```
[1] 0.1226
```

At around 0.12 the root mean squared error is slightly smaller than that of model `fit`. So, for this example the addition of holidays into the modeling framework has a marginal impact on overall performance. Oftentimes, you will find an even greater impact, especially for business time series data, where public holidays can have a significant impact on the demand for goods and services.

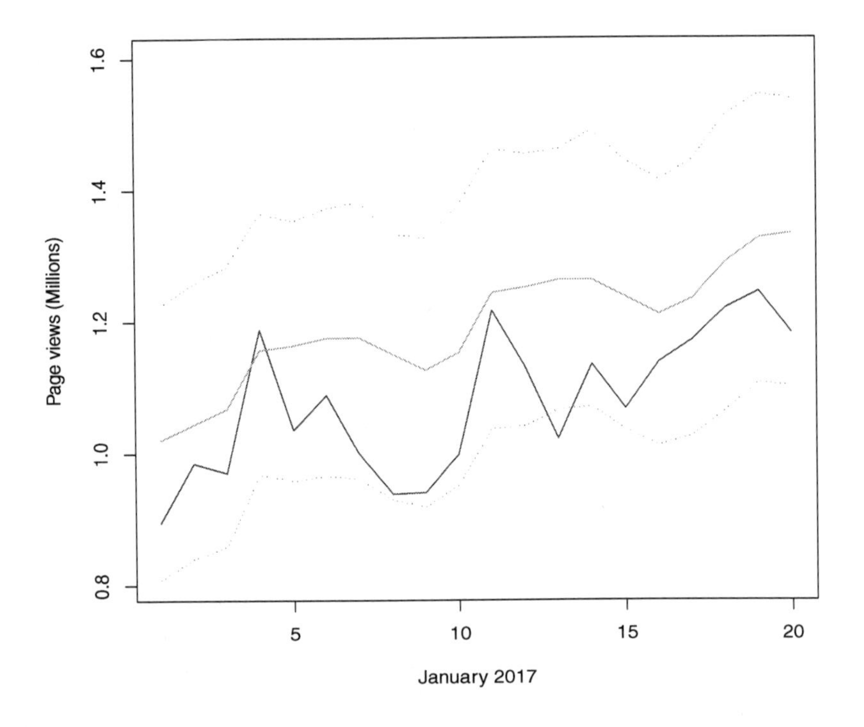

Figure 9.10: Predicted and actual values for `fit1`

Figure 9.10 depicts the actual values alongside the predicted values and the uncertainty interval. It is interesting to note that, as with the previous model, the predicted values are consistently higher (almost) than the actual observations. For the most part the actual observations lie comfortably within the uncertainty interval.

Limitations of the Prophet Framework

The algorithm came out of the desire to forecast business time series data collected and analyzed by Facebook. It is therefore not suitable for all types of time series observations, nevertheless as the procedure is tested in other domains, successful applications will no doubt emerge.

Summary

The Prophet time series forecasting framework is a powerful set of tools that can lead to highly accurate forecasts of future time series observations. It is extremely easy to use in R, and is very fast (even with large data samples). This makes it suitable for situations where there are large numbers of business time series which need to be predicted. It was designed to be used by anyone who needs to generate forecasts, no matter what their technical skill level may be, and it is robust to missing data, shifts in the trend, and outliers.

Chapter 10

Final Thoughts

IN this text we scratched the surface of time series forecasting in R. In fact, the subject of time series forecasting is so vast and so broad that perhaps we exaggerate about scratching the surface - more like a gentle glance into the possible. So, when you go from here? Assuming I haven't scared you off time series forecasting, there are a number of things you can do next.

Functions and Parameters

Every function in the models we discussed has a number of parameters. Throughout the book we tweaked and adjusted some of these parameters, which often yielded improvements in performance. Of course, in a textbook this size we can only cover a very limited number of tweaks and adjustments.

You are encouraged to try as many alternatives as you have time to experiment with. The fastest way to get up to speed with the parameters inherent in a specific R representation, is to read the associated documentation that comes with the package. They almost always provide snippets of examples and details of default or acceptable values.

For each of the models discussed in this text, grab a copy

of the associated package documentation and go straight to the functions we outlined, and study the additional parameters we did not discuss. Then, using the data you already have played with from this book explore what happens when you add in additional parameters, or change the values of preexisting parameters. Your experience, understanding and insight will grow exponentially if you do this.

New Data

Today, there is no shortage of data for which time series models can be applied. Public repositories of data are freely available over the Internet for government, business, and science. Seek out time series data sets in areas that you are interested in, and evaluate how well the models in this text perform on this data.

Another idea is to look for opportunities within the organization for which you work (or are interested in working in) to apply time series forecasting models. Managers and leaders are very interested in the new insights that can be brought to their attention by fresh analysis of preexisting data.

Volunteer to work in an existing project, or raise your hand and start your own. If you do this, you'll be amazed at how quickly coworkers and leaders will identify you as the expert in the area, and come to you seeking advice and analysis. Using this technique alone you will quickly master any of forecasting techniques discussed in this book from a practical and pragmatic perspective.

Review Applications

The time series forecasting techniques discussed in this book have been used successfully across a very broad range of disciplines. In your area of interest, keep your eye open for new applications of these models. Reports featuring the use of these models for very specific problems are frequently detailed in the practitioner and applied journals. Here are three sources where you can find such articles:

- `https://www.hindawi.com/`
- `https://www.plos.org/`
- `http://www.mdpi.com/`

Commit to reading one or two articles a week covering the specific model you are interested in. In no time at all, you will be a master of the technique.

Network

The old adage goes "it's not what you know, but who you know." I like to say, "*who you know will boost what you know*", for that reason joining a forecasting group or society can strengthen your network and improve your knowledge.

There are many such groups out there. Become a member and get involved. This can be one of the most cost-effective ways of meeting people further along the road than you. At the very least, you should consider joining a local R user group, and attend their meeting on a regular basis. If no groups exist near you, form one.

Congratulations!

You made it to the end. Here are three things you can do next.

1. Pick up your FREE copy of **12 Resources to Supercharge Your Productivity in R** at *http://www.auscov.com*

2. Gift a copy of this book to your friends, co-workers, teammates or your entire organization.

3. If you found this book useful and have a moment to spare, I would really appreciate a short review. Your help in spreading the word is gratefully received.

I've spoken to thousands of people over the past few years. I'd love to hear your experiences using the ideas in this book. Contact me with your stories, questions and suggestions at *Info@NigelDLewis.com*.

Dr. N.D. Lewis

Good luck!

P.S. Thanks for allowing me to partner with you on your data science journey.

Index

OTHER BOOKS YOU WILL ALSO ENJOY

- Machine Learning Made Easy with R
- Neural Networks for Time Series Forecasting with R
- Deep Learning Made Easy with R:
  - Volume I: A Gentle Introduction for Data Science
  - Volume II: Practical Tools for Data Science
  - Volume III: Breakthrough Techniques to Transform Performance
- Deep Learning for Business with R
- Build Your Own Neural Network TODAY!
- 92 Applied Predictive Modeling Techniques in R
- 100 Statistical Tests in R
- Visualizing Complex Data Using R
- Learning from Data Made Easy with R
- Deep Time Series Forecasting with Python
- Deep Learning for Business with Python
- Deep Learning Step by Step with Python

For further detail's visit www.AusCov.com

Write your notes here:

Made in the USA
Middletown, DE
27 March 2018